Excel

YEAR 7

The Complete Fractions Workbook

ESSENTIAL skills

Get the Results You Want!

PASCAL PRESS

AS Kalra

Reprinted 2001, 2003, 2004, 2006, 2007, 2009, 2010, 2011, 2013, 2014, 2016, 2017, 2019, 2020, 2021, 2022, 2023, 2024

ISBN 978 1 74020 043 1

Pascal Press
PO Box 250
Glebe NSW 2037
(02) 9198 1748
www.pascalpress.com.au

Publisher: Vivienne Joannou
Typeset by Big Picture Communications and Grizzly Graphics (Leanne Richters)
Cover by DiZign Pty Ltd
Printed by Vivar Printing/Green Giant Press

Dedication
This book is dedicated to the new generation of young Australians in whose hands lies the future of our nation and who by their hard work, acquired knowledge and intelligence will take Australia successfully through the 21st century.

This book is also in the loving, living and lasting memory of my dear mum, dad and uncle, who will remain a great source of inspiration and encouragement to me for times to come.

Acknowledgements
I would especially like to express my thanks and appreciation to my dear wife and my dear son, who have helped me to find the time to write this book. Without their help and support, achievement of all this work would not have been possible.

CONTENTS

INTRODUCTION

Mathematics is a highly important subject in our school syllabus. It is the study of numbers—which surround us everywhere in our lives.

This workbook of fractions for Year 7 is designed to make students feel confident in the basic processes of fractions. It will help to satisfy the needs of slower learners, and provide enrichment opportunities for quicker learners.

The important features of this book are:

☞ It is organised into seven sections. Great care has been taken with the development of each unit. Lots of examples are given so that students can go through each unit confidently. Many exercises are provided to give sufficient practice to students.

☞ There is a set of **Quick Questions** in every unit. Sufficient exercises of all standards are given to make sure that students think fast and remain in touch with the previous work.

☞ **Stop. Revise. Check.** is a complete summary of the main concepts covered in the chapter. Each concept is explained with a worked example to ensure that students have fully understood the material. Students should read this page thoroughly to prepare themselves for the test; they should be 100% confident that they know their work before attempting the test. If there are any concepts students are unsure of, they should go back and revise.

☞ **Stay In Touch** is intended to keep students in touch with what they have done before. It ensures that all topics receive constant revision.

☞ A mixed test at the end of each section, and four review exams at the end of the book, have been included so that students can test their knowledge of the basic skills of fractions. Each test paper has a marking scheme so that students can mark their own answers. This should give them a fair idea of how much they have achieved.

Considerable effort has been made to create an interesting, appealing and meaningful book that students will find valuable and enjoyable.

My best wishes are with you and I believe that this book will help you achieve the best possible results. Good luck in your studies!

AS Kalra

SECTION 1

UNIT 1 Types of Fractions

There are three different types of fractions.

1. **Proper fractions:** A proper fraction is a fraction in which the numerator is less than the denominator.

   For example $\frac{2}{3}$, $\frac{4}{5}$, $\frac{1}{7}$, $\frac{5}{8}$, $\frac{3}{11}$

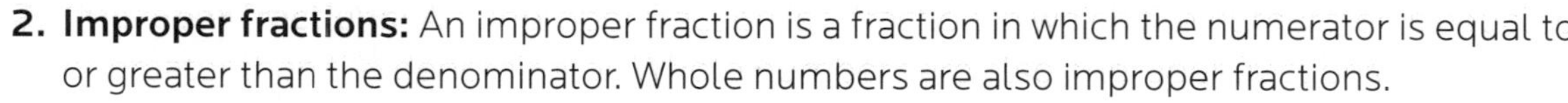

2. **Improper fractions:** An improper fraction is a fraction in which the numerator is equal to or greater than the denominator. Whole numbers are also improper fractions.

   For example $\frac{4}{3}$, $\frac{9}{5}$, $\frac{7}{2}$, $\frac{19}{4}$, $\frac{61}{7}$, $\frac{10}{1}$ (10), $\frac{12}{3}$ (4), $\frac{5}{5}$ (1)

3. **Mixed numbers:** A mixed number has a whole number and a proper fraction part.

   For example $3\frac{2}{5}$, $5\frac{1}{3}$, $9\frac{2}{7}$, $6\frac{9}{11}$, $7\frac{1}{9}$

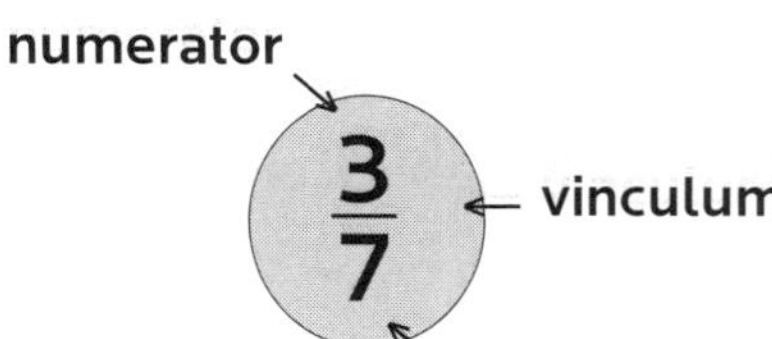

State whether the following fractions are **proper** (P), **improper** (I) or **mixed numbers** (MN).

1 $\frac{2}{3}$ ________	**2** $\frac{5}{4}$ ________	**3** $\frac{6}{7}$ ________	**4** $\frac{8}{3}$ ________
5 $2\frac{1}{4}$ ________	**6** $\frac{5}{11}$ ________	**7** $3\frac{1}{8}$ ________	**8** $\frac{6}{19}$ ________
9 $\frac{2}{7}$ ________	**10** $\frac{8}{5}$ ________	**11** $3\frac{2}{5}$ ________	**12** $6\frac{2}{3}$ ________
13 8 ________	**14** 3 ________	**15** $\frac{5}{13}$ ________	**16** $1\frac{1}{2}$ ________
17 $\frac{5}{9}$ ________	**18** $\frac{9}{23}$ ________	**19** $4\frac{5}{8}$ ________	**20** $3\frac{1}{2}$ ________
21 $2\frac{1}{7}$ ________	**22** $\frac{9}{2}$ ________	**23** $\frac{6}{13}$ ________	**24** $\frac{12}{5}$ ________
25 $\frac{2}{13}$ ________	**26** $3\frac{3}{4}$ ________	**27** $5\frac{1}{2}$ ________	**28** $\frac{6}{17}$ ________
29 $\frac{9}{10}$ ________	**30** 12 ________	**31** $6\frac{1}{3}$ ________	**32** 5 ________
33 $6\frac{2}{5}$ ________	**34** $\frac{3}{16}$ ________	**35** $\frac{18}{7}$ ________	**36** $\frac{23}{5}$ ________
37 $6\frac{5}{6}$ ________	**38** $\frac{3}{8}$ ________	**39** 7 ________	**40** $\frac{21}{13}$ ________
41 $3\frac{2}{9}$ ________	**42** $\frac{15}{7}$ ________	**43** $\frac{8}{27}$ ________	**44** $5\frac{2}{11}$ ________

UNIT 2 Changing Improper Fractions to Mixed Numbers

Improper fractions can be easily changed to mixed numbers.

Example: Change the following improper fractions to mixed numbers.

(a) $\frac{9}{4}$ (b) $\frac{12}{5}$ (c) $\frac{19}{3}$ (d) $\frac{65}{7}$

Solution:

(a) $\frac{9}{4} = 9 \div 4$
$= 2\frac{1}{4}$

(b) $\frac{12}{5} = 12 \div 5$
$= 2\frac{2}{5}$

(c) $\frac{19}{3} = 19 \div 3$
$= 6\frac{1}{3}$

(d) $\frac{65}{7} = 65 \div 7$
$= 9\frac{2}{7}$

Change the following improper fractions to mixed numbers.

1 $\frac{3}{2}$ = ________ **2** $\frac{5}{2}$ = ________ **3** $\frac{7}{2}$ = ________ **4** $\frac{9}{2}$ = ________

5 $\frac{4}{3}$ = ________ **6** $\frac{5}{3}$ = ________ **7** $\frac{7}{3}$ = ________ **8** $\frac{11}{3}$ = ________

9 $\frac{7}{4}$ = ________ **10** $\frac{9}{4}$ = ________ **11** $\frac{11}{4}$ = ________ **12** $\frac{13}{4}$ = ________

13 $\frac{8}{5}$ = ________ **14** $\frac{9}{5}$ = ________ **15** $\frac{11}{5}$ = ________ **16** $\frac{13}{5}$ = ________

17 $\frac{25}{2}$ = ________ **18** $\frac{19}{5}$ = ________ **19** $\frac{27}{4}$ = ________ **20** $\frac{34}{5}$ = ________

21 $\frac{13}{6}$ = ________ **22** $\frac{28}{3}$ = ________ **23** $\frac{19}{9}$ = ________ **24** $\frac{17}{5}$ = ________

25 $\frac{36}{7}$ = ________ **26** $\frac{39}{9}$ = ________ **27** $\frac{28}{5}$ = ________ **28** $\frac{37}{5}$ = ________

29 $\frac{61}{8}$ = ________ **30** $\frac{35}{9}$ = ________ **31** $\frac{57}{6}$ = ________ **32** $\frac{75}{7}$ = ________

33 $\frac{32}{7}$ = ________ **34** $\frac{83}{12}$ = ________ **35** $\frac{65}{9}$ = ________ **36** $\frac{48}{11}$ = ________

37 $\frac{89}{11}$ = ________ **38** $\frac{93}{6}$ = ________ **39** $\frac{86}{9}$ = ________ **40** $\frac{97}{12}$ = ________

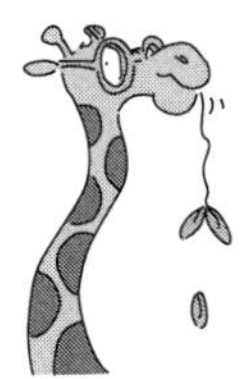

In a fraction
the top part is called the numerator
and the bottom part
is called the denominator.

UNIT 3 Changing Mixed Numbers to Improper Fractions

Mixed numbers can be easily changed to improper fractions.

Example: Change the following mixed numbers to improper fractions.

(a) $1\frac{2}{3}$ (b) $2\frac{5}{7}$ (c) $5\frac{3}{4}$

Solution:

(a) $1\frac{2}{3} = \frac{3}{3} + \frac{2}{3} = \frac{3+2}{3} = \frac{5}{3}$

(b) $2\frac{5}{7} = \frac{14}{7} + \frac{5}{7} = \frac{14+5}{7} = \frac{19}{7}$

(c) $5\frac{3}{4} = \frac{20}{4} + \frac{3}{4} = \frac{20+3}{4} = \frac{23}{4}$

But there is a quicker way to work it out.
Multiply the denominator with the whole number and add the numerator.

(a) $1\frac{2}{3} = \frac{3 \times 1 + 2}{3} = \frac{3+2}{3} = \frac{5}{3}$

(b) $2\frac{5}{7} = \frac{7 \times 2 + 5}{7} = \frac{14+5}{7} = \frac{19}{7}$

(c) $5\frac{3}{4} = \frac{4 \times 5 + 3}{4} = \frac{20+3}{4} = \frac{23}{4}$

Change the following mixed numbers to fractions.

1 $1\frac{1}{4}$ = ______ **2** $1\frac{1}{5}$ = ______ **3** $1\frac{1}{3}$ = ______ **4** $1\frac{2}{7}$ = ______

5 $2\frac{1}{2}$ = ______ **6** $3\frac{1}{2}$ = ______ **7** $4\frac{1}{2}$ = ______ **8** $5\frac{1}{2}$ = ______

9 $2\frac{1}{3}$ = ______ **10** $3\frac{1}{3}$ = ______ **11** $4\frac{1}{3}$ = ______ **12** $5\frac{1}{3}$ = ______

13 $6\frac{1}{4}$ = ______ **14** $7\frac{1}{4}$ = ______ **15** $8\frac{1}{4}$ = ______ **16** $9\frac{1}{4}$ = ______

17 $3\frac{2}{7}$ = ______ **18** $5\frac{1}{4}$ = ______ **19** $8\frac{1}{5}$ = ______ **20** $6\frac{1}{9}$ = ______

21 $8\frac{3}{4}$ = ______ **22** $12\frac{3}{5}$ = ______ **23** $17\frac{1}{2}$ = ______ **24** $23\frac{2}{3}$ = ______

25 $13\frac{2}{3}$ = ______ **26** $18\frac{2}{5}$ = ______ **27** $24\frac{1}{3}$ = ______ **28** $22\frac{1}{4}$ = ______

29 $21\frac{3}{4}$ = ______ **30** $16\frac{2}{3}$ = ______ **31** $18\frac{2}{3}$ = ______ **32** $15\frac{5}{6}$ = ______

33 $14\frac{2}{3}$ = ______ **34** $11\frac{1}{5}$ = ______ **35** $25\frac{3}{4}$ = ______ **36** $26\frac{1}{2}$ = ______

UNIT 4 Simplifying Fractions (with Highest Common Factor 2 or 5)

A fraction is said to be in its simplest form when its numerator and denominator are as small as possible.

To write the fraction in its simplest form, divide both the numerator and denominator by the highest common factor (H.C.F.).

Example: Write the following fractions in their simplest form.

(a) $\frac{4}{6}$ (b) $\frac{10}{12}$ (c) $\frac{5}{10}$ (d) $\frac{15}{20}$

Solution:

(a) $\frac{4}{6} = \frac{2}{3}$ (÷2) (b) $\frac{10}{12} = \frac{5}{6}$ (÷2) (c) $\frac{5}{10} = \frac{1}{2}$ (÷5) (d) $\frac{15}{20} = \frac{3}{4}$ (÷5)

A proper fraction is when the numerator is smaller than the denominator.

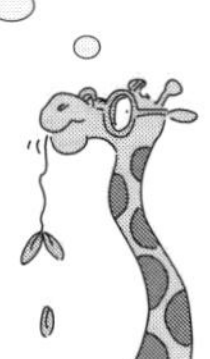

Write the following fractions in their simplest form.

1 $\frac{2}{4}$ = ______	**2** $\frac{6}{8}$ = ______	**3** $\frac{8}{10}$ = ______	**4** $\frac{12}{14}$ = ______
5 $\frac{2}{6}$ = ______	**6** $\frac{4}{10}$ = ______	**7** $\frac{6}{14}$ = ______	**8** $\frac{10}{14}$ = ______
9 $\frac{2}{8}$ = ______	**10** $\frac{2}{12}$ = ______	**11** $\frac{4}{14}$ = ______	**12** $\frac{6}{16}$ = ______
13 $\frac{8}{18}$ = ______	**14** $\frac{10}{24}$ = ______	**15** $\frac{12}{22}$ = ______	**16** $\frac{14}{26}$ = ______
17 $\frac{16}{22}$ = ______	**18** $\frac{20}{26}$ = ______	**19** $\frac{28}{30}$ = ______	**20** $\frac{42}{50}$ = ______
21 $\frac{5}{10}$ = ______	**22** $\frac{5}{15}$ = ______	**23** $\frac{5}{20}$ = ______	**24** $\frac{5}{25}$ = ______
25 $\frac{10}{15}$ = ______	**26** $\frac{20}{25}$ = ______	**27** $\frac{30}{35}$ = ______	**28** $\frac{40}{45}$ = ______
29 $\frac{5}{40}$ = ______	**30** $\frac{10}{25}$ = ______	**31** $\frac{15}{35}$ = ______	**32** $\frac{20}{45}$ = ______
33 $\frac{25}{30}$ = ______	**34** $\frac{30}{55}$ = ______	**35** $\frac{35}{40}$ = ______	**36** $\frac{40}{65}$ = ______
37 $\frac{45}{50}$ = ______	**38** $\frac{55}{60}$ = ______	**39** $\frac{65}{75}$ = ______	**40** $\frac{75}{85}$ = ______

UNIT 5 Simplifying Fractions (with Highest Common Factor 10)

A fraction is in its simplest form when its numerator and denominator are divided by their highest common factor.

Example: Write the following fractions in their simplest form.

(a) $\frac{10}{20}$ (b) $\frac{20}{30}$ (c) $\frac{10}{50}$ (d) $\frac{30}{70}$

Solution:

(a) $\frac{10}{20} \overset{\div 10}{=} \frac{1}{2}$ (b) $\frac{20}{30} \overset{\div 10}{=} \frac{2}{3}$ (c) $\frac{10}{50} \overset{\div 10}{=} \frac{1}{5}$ (d) $\frac{30}{70} \overset{\div 10}{=} \frac{3}{7}$

Write the following fractions in their simplest form.

1 $\frac{10}{30}$ = ______	**2** $\frac{10}{40}$ = ______	**3** $\frac{10}{60}$ = ______	**4** $\frac{10}{70}$ = ______
5 $\frac{20}{50}$ = ______	**6** $\frac{20}{70}$ = ______	**7** $\frac{20}{90}$ = ______	**8** $\frac{30}{40}$ = ______
9 $\frac{30}{50}$ = ______	**10** $\frac{30}{80}$ = ______	**11** $\frac{30}{100}$ = ______	**12** $\frac{30}{110}$ = ______
13 $\frac{10}{80}$ = ______	**14** $\frac{20}{110}$ = ______	**15** $\frac{30}{130}$ = ______	**16** $\frac{40}{50}$ = ______
17 $\frac{40}{70}$ = ______	**18** $\frac{40}{90}$ = ______	**19** $\frac{40}{110}$ = ______	**20** $\frac{40}{130}$ = ______
21 $\frac{50}{60}$ = ______	**22** $\frac{50}{70}$ = ______	**23** $\frac{50}{80}$ = ______	**24** $\frac{50}{90}$ = ______
25 $\frac{60}{70}$ = ______	**26** $\frac{60}{110}$ = ______	**27** $\frac{60}{130}$ = ______	**28** $\frac{60}{170}$ = ______
29 $\frac{70}{80}$ = ______	**30** $\frac{70}{90}$ = ______	**31** $\frac{70}{100}$ = ______	**32** $\frac{70}{110}$ = ______
33 $\frac{80}{90}$ = ______	**34** $\frac{80}{110}$ = ______	**35** $\frac{80}{130}$ = ______	**36** $\frac{80}{150}$ = ______
37 $\frac{90}{100}$ = ______	**38** $\frac{90}{110}$ = ______	**39** $\frac{90}{130}$ = ______	**40** $\frac{90}{140}$ = ______
41 $\frac{100}{110}$ = ______	**42** $\frac{100}{130}$ = ______	**43** $\frac{100}{170}$ = ______	**44** $\frac{110}{120}$ = ______

UNIT 6 Simplifying Fractions
(with Highest Common Factor an Odd Number)

A fraction can be reduced to its simplest form if we divide both the numerator and the denominator by their highest common factor.

A proper fraction is in its simplest form when its numerator and denominator are as small as possible.

Example: Write the following fractions in their simplest form.

(a) $\frac{3}{6}$ (b) $\frac{7}{14}$ (c) $\frac{9}{27}$ (d) $\frac{11}{33}$

Solution:

(a) $\frac{3}{6} = \frac{1}{2}$ (÷3) (b) $\frac{7}{14} = \frac{1}{2}$ (÷7) (c) $\frac{9}{27} = \frac{1}{3}$ (÷9) (d) $\frac{11}{33} = \frac{1}{3}$ (÷11)

Write the following fractions in their simplest form.

1 $\frac{3}{9}$ = ______	**2** $\frac{3}{12}$ = ______	**3** $\frac{3}{15}$ = ______	**4** $\frac{3}{18}$ = ______
5 $\frac{6}{9}$ = ______	**6** $\frac{9}{12}$ = ______	**7** $\frac{9}{15}$ = ______	**8** $\frac{12}{21}$ = ______
9 $\frac{7}{21}$ = ______	**10** $\frac{7}{28}$ = ______	**11** $\frac{7}{35}$ = ______	**12** $\frac{7}{42}$ = ______
13 $\frac{14}{21}$ = ______	**14** $\frac{21}{35}$ = ______	**15** $\frac{21}{28}$ = ______	**16** $\frac{21}{49}$ = ______
17 $\frac{9}{18}$ = ______	**18** $\frac{9}{27}$ = ______	**19** $\frac{9}{36}$ = ______	**20** $\frac{9}{45}$ = ______
21 $\frac{18}{27}$ = ______	**22** $\frac{27}{36}$ = ______	**23** $\frac{36}{45}$ = ______	**24** $\frac{45}{54}$ = ______
25 $\frac{11}{22}$ = ______	**26** $\frac{11}{33}$ = ______	**27** $\frac{11}{44}$ = ______	**28** $\frac{11}{55}$ = ______
29 $\frac{22}{55}$ = ______	**30** $\frac{33}{55}$ = ______	**31** $\frac{33}{77}$ = ______	**32** $\frac{22}{99}$ = ______
33 $\frac{15}{35}$ = ______	**34** $\frac{25}{55}$ = ______	**35** $\frac{35}{75}$ = ______	**36** $\frac{15}{85}$ = ______
37 $\frac{12}{63}$ = ______	**38** $\frac{21}{77}$ = ______	**39** $\frac{27}{63}$ = ______	**40** $\frac{66}{77}$ = ______
41 $\frac{42}{49}$ = ______	**42** $\frac{54}{63}$ = ______	**43** $\frac{99}{110}$ = ______	**44** $\frac{30}{75}$ = ______

Quick Questions

This unit of 'Quick Questions' is designed to make sure that you think fast and remain in touch with the previous work of the section.

A mixed number has a whole number and a proper fraction part.

Change the following improper fractions to mixed numbers.

1 $\frac{3}{2}$ = ________ **2** $\frac{5}{4}$ = ________ **3** $\frac{4}{3}$ = ________ **4** $\frac{5}{3}$ = ________

5 $\frac{5}{2}$ = ________ **6** $\frac{6}{5}$ = ________ **7** $\frac{7}{2}$ = ________ **8** $\frac{7}{3}$ = ________

9 $\frac{7}{4}$ = ________ **10** $\frac{7}{5}$ = ________ **11** $\frac{7}{6}$ = ________ **12** $\frac{8}{3}$ = ________

13 $\frac{8}{5}$ = ________ **14** $\frac{8}{7}$ = ________ **15** $\frac{9}{2}$ = ________ **16** $\frac{9}{4}$ = ________

Change the following mixed numbers to improper fractions.

17 $1\frac{1}{2}$ = ________ **18** $2\frac{1}{2}$ = ________ **19** $3\frac{1}{2}$ = ________ **20** $4\frac{1}{2}$ = ________

21 $1\frac{1}{4}$ = ________ **22** $2\frac{1}{4}$ = ________ **23** $3\frac{1}{4}$ = ________ **24** $4\frac{1}{4}$ = ________

25 $2\frac{2}{3}$ = ________ **26** $2\frac{2}{5}$ = ________ **27** $2\frac{2}{7}$ = ________ **28** $2\frac{2}{9}$ = ________

29 $3\frac{1}{3}$ = ________ **30** $4\frac{2}{3}$ = ________ **31** $5\frac{2}{5}$ = ________ **32** $6\frac{1}{2}$ = ________

Simplify the following fractions.

33 $\frac{2}{4}$ = ________ **34** $\frac{2}{6}$ = ________ **35** $\frac{2}{8}$ = ________ **36** $\frac{2}{10}$ = ________

37 $\frac{3}{6}$ = ________ **38** $\frac{6}{9}$ = ________ **39** $\frac{9}{12}$ = ________ **40** $\frac{12}{15}$ = ________

41 $\frac{5}{10}$ = ________ **42** $\frac{5}{25}$ = ________ **43** $\frac{5}{35}$ = ________ **44** $\frac{5}{55}$ = ________

45 $\frac{10}{20}$ = ________ **46** $\frac{10}{30}$ = ________ **47** $\frac{20}{30}$ = ________ **48** $\frac{30}{40}$ = ________

UNIT 8

Stay in Touch

The unit 'Stay In Touch' is intended to keep you in touch with what you have done before.

It also ensures that all types of questions in this section receive constant revision.

State whether the following fractions are **proper** (P), **improper** (I) or **mixed numbers** (MN).

1 $\frac{3}{7}$ ______	**2** $\frac{9}{8}$ ______	**3** $2\frac{1}{2}$ ______	**4** $5\frac{1}{6}$ ______
5 $\frac{2}{5}$ ______	**6** 11 ______	**7** $\frac{18}{7}$ ______	**8** $\frac{23}{19}$ ______
9 5 ______	**10** $7\frac{1}{2}$ ______	**11** $\frac{12}{31}$ ______	**12** $\frac{45}{29}$ ______

Change the following improper fractions to mixed numbers.

13 $\frac{19}{4}$ = ______	**14** $\frac{21}{5}$ = ______	**15** $\frac{36}{7}$ = ______	**16** $\frac{56}{9}$ = ______
17 $\frac{38}{7}$ = ______	**18** $\frac{62}{9}$ = ______	**19** $\frac{25}{8}$ = ______	**20** $\frac{33}{8}$ = ______
21 $\frac{19}{2}$ = ______	**22** $\frac{63}{8}$ = ______	**23** $\frac{51}{9}$ = ______	**24** $\frac{43}{8}$ = ______

Change the following mixed numbers to improper fractions.

25 $11\frac{2}{3}$ = ______	**26** $15\frac{1}{4}$ = ______	**27** $16\frac{2}{3}$ = ______	**28** $12\frac{3}{4}$ = ______
29 $8\frac{5}{8}$ = ______	**30** $9\frac{2}{3}$ = ______	**31** $6\frac{7}{8}$ = ______	**32** $15\frac{2}{3}$ = ______
33 $16\frac{1}{2}$ = ______	**34** $13\frac{1}{4}$ = ______	**35** $23\frac{1}{4}$ = ______	**36** $51\frac{1}{2}$ = ______

Simplify the following fractions.

37 $\frac{6}{28}$ = ______	**38** $\frac{8}{26}$ = ______	**39** $\frac{15}{35}$ = ______	**40** $\frac{16}{40}$ = ______
41 $\frac{12}{42}$ = ______	**42** $\frac{18}{24}$ = ______	**43** $\frac{28}{49}$ = ______	**44** $\frac{32}{56}$ = ______
45 $\frac{26}{39}$ = ______	**46** $\frac{27}{72}$ = ______	**47** $\frac{24}{60}$ = ______	**48** $\frac{30}{110}$ = ______

UNIT 9 Stop. Revise. Check.

☞ A fraction = $\frac{\text{Numerator}}{\text{Denominator}}$ e.g. $\frac{2}{3}$

☞ A proper fraction is a fraction in which the numerator is less than the denominator.

e.g. $\frac{4}{7}, \frac{3}{11}, \frac{35}{43}$

☞ An improper fraction is a fraction in which the numerator is equal to or greater than the denominator.

e.g. $\frac{7}{7}, \frac{9}{2}, \frac{21}{5}, \frac{12}{3}, \frac{8}{1}$

☞ A mixed number has a whole number and a proper fraction part.

e.g. $3\frac{1}{2}, 7\frac{3}{4}, 5\frac{7}{33}$

☞ **Example:** Change the following improper fractions to mixed numbers.

(a) $\frac{9}{2}$ (b) $\frac{13}{5}$ (c) $\frac{48}{7}$

Solution:

(a) $\frac{9}{2} = 9 \div 2$
$= 4\frac{1}{2}$

(b) $\frac{13}{5} = 13 \div 5$
$= 2\frac{3}{5}$

(c) $\frac{48}{7} = 48 \div 7$
$= 6\frac{6}{7}$

☞ **Example:** Change the following mixed numbers to improper fractions.

(a) $2\frac{3}{5}$ (b) $3\frac{1}{2}$ (c) $5\frac{1}{4}$

Solution:

(a) $2\frac{3}{5} = \frac{5 \times 2 + 3}{5}$
$= \frac{13}{5}$

(b) $3\frac{1}{2} = \frac{2 \times 3 + 1}{2}$
$= \frac{7}{2}$

(c) $5\frac{1}{4} = \frac{4 \times 5 + 1}{4}$
$= \frac{21}{4}$

☞ To write a fraction in its simplest form, divide both the numerator and denominator by the highest common factor (H.C.F.).

Example: Simplify the following fractions.

(a) $\frac{4}{12}$ (b) $\frac{15}{20}$ (c) $\frac{20}{50}$ (d) $\frac{21}{35}$

Solution:

(a) $\frac{4}{12} = \frac{4 \div 4}{12 \div 4}$
$= \frac{1}{3}$

(b) $\frac{15}{20} = \frac{15 \div 5}{20 \div 5}$
$= \frac{3}{4}$

(c) $\frac{20}{50} = \frac{20 \div 10}{50 \div 10}$
$= \frac{2}{5}$

(d) $\frac{21}{35} = \frac{21 \div 7}{35 \div 7}$
$= \frac{3}{5}$

UNIT 10 Test on Section 1

Instructions: Do not use a calculator.
Attempt all questions.

Time allowed: 15 minutes **Total marks:** 48

Marks

Question 1: State whether the following fractions are **proper** (P), **improper** (I) or a **mixed number** (MN).

				Marks
a $\frac{9}{5}$ ______	**b** $\frac{23}{16}$ ______	**c** $\frac{5}{12}$ ______	**d** 6 ______	4
e $2\frac{3}{4}$ ______	**f** $\frac{51}{7}$ ______	**g** $\frac{9}{35}$ ______	**h** $8\frac{6}{11}$ ______	4
i 9 ______	**j** $\frac{63}{5}$ ______	**k** $\frac{2}{17}$ ______	**l** $6\frac{1}{4}$ ______	4

Question 2: Change the following improper fractions to mixed numbers.

				Marks
a $\frac{18}{5}$ = ______	**b** $\frac{64}{7}$ = ______	**c** $\frac{32}{9}$ = ______	**d** $\frac{100}{11}$ = ______	4
e $\frac{43}{6}$ = ______	**f** $\frac{57}{8}$ = ______	**g** $\frac{41}{3}$ = ______	**h** $\frac{37}{5}$ = ______	4
i $\frac{23}{2}$ = ______	**j** $\frac{61}{9}$ = ______	**k** $\frac{53}{10}$ = ______	**l** $\frac{38}{7}$ = ______	4

Question 3: Change the following mixed numbers to improper fractions.

				Marks
a $5\frac{3}{4}$ = ______	**b** $8\frac{1}{2}$ = ______	**c** $10\frac{3}{7}$ = ______	**d** $5\frac{4}{5}$ = ______	4
e $3\frac{6}{7}$ = ______	**f** $9\frac{4}{5}$ = ______	**g** $10\frac{2}{3}$ = ______	**h** $12\frac{1}{3}$ = ______	4
i $15\frac{5}{6}$ = ______	**j** $12\frac{3}{4}$ = ______	**k** $11\frac{2}{3}$ = ______	**l** $16\frac{2}{5}$ = ______	4

Question 4: Simplify the following fractions.

				Marks
a $\frac{8}{30}$ = ______	**b** $\frac{12}{32}$ = ______	**c** $\frac{15}{65}$ = ______	**d** $\frac{18}{52}$ = ______	4
e $\frac{25}{90}$ = ______	**f** $\frac{60}{80}$ = ______	**g** $\frac{120}{150}$ = ______	**h** $\frac{28}{64}$ = ______	4
i $\frac{36}{80}$ = ______	**j** $\frac{24}{88}$ = ______	**k** $\frac{32}{72}$ = ______	**l** $\frac{27}{81}$ = ______	4

Total marks = ___ / 48

SECTION 2

UNIT 11 Equivalent Fractions

Equivalent fractions have the same value and they represent the same number but they have different numerators and denominators. For example, $\frac{1}{3} = \frac{2}{6} = \frac{3}{9}$ have the same value even though they look different.

Example: Complete the following equivalent fractions.

(a) $\frac{1}{2} = \frac{2}{\square}$ (b) $\frac{1}{3} = \frac{4}{\square}$ (c) $\frac{2}{5} = \frac{6}{\square}$ (d) $\frac{3}{4} = \frac{18}{\square}$

Solution:

(a) $\frac{1 \times 2}{2 \times 2} = \frac{2}{4}$ (b) $\frac{1 \times 4}{3 \times 4} = \frac{4}{12}$ (c) $\frac{2 \times 3}{5 \times 3} = \frac{6}{15}$ (d) $\frac{3 \times 6}{4 \times 6} = \frac{18}{24}$

Complete the following equivalent fractions.

1 $\frac{1}{2} = \frac{3}{\square}$ **2** $\frac{1}{3} = \frac{2}{\square}$ **3** $\frac{1}{4} = \frac{5}{\square}$ **4** $\frac{1}{5} = \frac{3}{\square}$

5 $\frac{2}{3} = \frac{6}{\square}$ **6** $\frac{2}{5} = \frac{10}{\square}$ **7** $\frac{2}{3} = \frac{8}{\square}$ **8** $\frac{2}{5} = \frac{12}{\square}$

9 $\frac{3}{4} = \frac{6}{\square}$ **10** $\frac{3}{5} = \frac{9}{\square}$ **11** $\frac{3}{7} = \frac{12}{\square}$ **12** $\frac{3}{8} = \frac{6}{\square}$

13 $\frac{1}{2} = \frac{7}{\square}$ **14** $\frac{2}{3} = \frac{12}{\square}$ **15** $\frac{4}{5} = \frac{20}{\square}$ **16** $\frac{6}{7} = \frac{42}{\square}$

17 $\frac{8}{9} = \frac{24}{\square}$ **18** $\frac{5}{7} = \frac{25}{\square}$ **19** $\frac{3}{7} = \frac{27}{\square}$ **20** $\frac{2}{9} = \frac{16}{\square}$

21 $\frac{3}{10} = \frac{12}{\square}$ **22** $\frac{2}{9} = \frac{10}{\square}$ **23** $\frac{5}{9} = \frac{15}{\square}$ **24** $\frac{3}{8} = \frac{18}{\square}$

25 $\frac{1}{2} = \frac{11}{\square}$ **26** $\frac{1}{3} = \frac{12}{\square}$ **27** $\frac{1}{4} = \frac{14}{\square}$ **28** $\frac{1}{5} = \frac{15}{\square}$

29 $\frac{3}{7} = \frac{12}{\square}$ **30** $\frac{4}{9} = \frac{12}{\square}$ **31** $\frac{5}{11} = \frac{20}{\square}$ **32** $\frac{6}{7} = \frac{30}{\square}$

33 $\frac{2}{9} = \frac{10}{\square}$ **34** $\frac{3}{7} = \frac{18}{\square}$ **35** $\frac{5}{13} = \frac{35}{\square}$ **36** $\frac{6}{11} = \frac{18}{\square}$

37 $\frac{3}{13} = \frac{12}{\square}$ **38** $\frac{7}{15} = \frac{35}{\square}$ **39** $\frac{2}{19} = \frac{6}{\square}$ **40** $\frac{13}{15} = \frac{39}{\square}$

UNIT 12 Equivalent Fractions

In order to obtain equivalent fractions, multiply or divide the numerator and denominator by the same number.

The bottom part of the fraction is called the denominator.

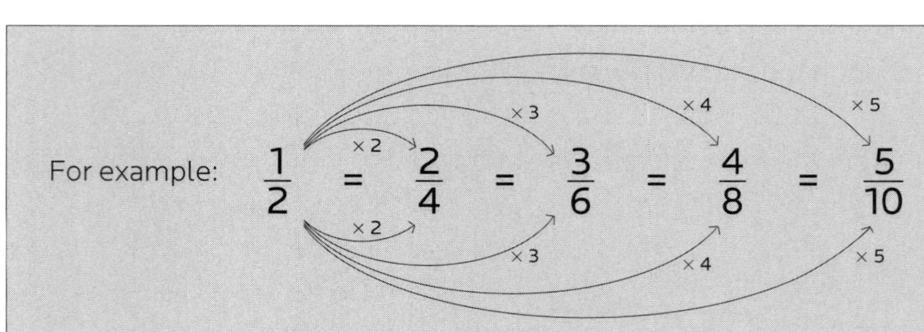

Complete the following equivalent fractions.

1 $\frac{1}{2} = \frac{\square}{4}$	**2** $\frac{1}{2} = \frac{\square}{6}$	**3** $\frac{1}{2} = \frac{\square}{8}$	**4** $\frac{1}{2} = \frac{\square}{10}$
5 $\frac{1}{3} = \frac{\square}{6}$	**6** $\frac{1}{3} = \frac{\square}{9}$	**7** $\frac{1}{3} = \frac{\square}{12}$	**8** $\frac{1}{3} = \frac{\square}{15}$
9 $\frac{2}{3} = \frac{\square}{18}$	**10** $\frac{1}{4} = \frac{\square}{24}$	**11** $\frac{2}{5} = \frac{\square}{30}$	**12** $\frac{1}{5} = \frac{\square}{30}$
13 $\frac{3}{4} = \frac{\square}{8}$	**14** $\frac{3}{5} = \frac{\square}{15}$	**15** $\frac{3}{4} = \frac{\square}{16}$	**16** $\frac{3}{5} = \frac{\square}{25}$
17 $\frac{2}{7} = \frac{\square}{21}$	**18** $\frac{3}{7} = \frac{\square}{28}$	**19** $\frac{3}{5} = \frac{\square}{35}$	**20** $\frac{3}{10} = \frac{\square}{30}$
21 $\frac{2}{3} = \frac{\square}{21}$	**22** $\frac{1}{4} = \frac{\square}{32}$	**23** $\frac{1}{2} = \frac{\square}{14}$	**24** $\frac{2}{3} = \frac{\square}{27}$
25 $\frac{1}{5} = \frac{\square}{10}$	**26** $\frac{1}{7} = \frac{\square}{14}$	**27** $\frac{2}{7} = \frac{\square}{21}$	**28** $\frac{2}{11} = \frac{\square}{22}$
29 $\frac{3}{5} = \frac{\square}{20}$	**30** $\frac{3}{7} = \frac{\square}{35}$	**31** $\frac{3}{5} = \frac{\square}{35}$	**32** $\frac{3}{8} = \frac{\square}{24}$
33 $\frac{4}{5} = \frac{\square}{20}$	**34** $\frac{4}{7} = \frac{\square}{42}$	**35** $\frac{4}{5} = \frac{\square}{25}$	**36** $\frac{4}{7} = \frac{\square}{49}$
37 $\frac{7}{8} = \frac{\square}{16}$	**38** $\frac{6}{7} = \frac{\square}{63}$	**39** $\frac{5}{6} = \frac{\square}{48}$	**40** $\frac{3}{5} = \frac{\square}{55}$

UNIT 13 Equivalent Fractions

Equivalent fractions represent the same number but have different numerators and denominators. It is very easy to form an equivalent fraction. All we have to do is multiply or divide the numerator and denominator by the same number.

Example: Complete the following equivalent fractions.

(a) $\frac{3}{6} = \frac{1}{\square}$ (b) $\frac{4}{12} = \frac{1}{\square}$ (c) $\frac{6}{15} = \frac{2}{\square}$ (d) $\frac{15}{25} = \frac{3}{\square}$

Solution:

(a) $\frac{3 \div 3}{6 \div 3} = \frac{1}{2}$ (b) $\frac{4 \div 4}{12 \div 4} = \frac{1}{3}$ (c) $\frac{6 \div 3}{15 \div 3} = \frac{2}{5}$ (d) $\frac{15 \div 5}{25 \div 5} = \frac{3}{5}$

Complete the following equivalent fractions.

1 $\frac{4}{6} = \frac{2}{\square}$ **2** $\frac{8}{12} = \frac{2}{\square}$ **3** $\frac{12}{16} = \frac{3}{\square}$ **4** $\frac{18}{24} = \frac{3}{\square}$

5 $\frac{9}{12} = \frac{3}{\square}$ **6** $\frac{20}{25} = \frac{4}{\square}$ **7** $\frac{7}{21} = \frac{1}{\square}$ **8** $\frac{5}{30} = \frac{1}{\square}$

9 $\frac{6}{16} = \frac{3}{\square}$ **10** $\frac{8}{18} = \frac{4}{\square}$ **11** $\frac{9}{36} = \frac{1}{\square}$ **12** $\frac{6}{36} = \frac{1}{\square}$

13 $\frac{12}{40} = \frac{3}{\square}$ **14** $\frac{10}{30} = \frac{1}{\square}$ **15** $\frac{3}{18} = \frac{1}{\square}$ **16** $\frac{5}{55} = \frac{1}{\square}$

17 $\frac{18}{45} = \frac{2}{\square}$ **18** $\frac{21}{28} = \frac{3}{\square}$ **19** $\frac{26}{32} = \frac{13}{\square}$ **20** $\frac{14}{21} = \frac{2}{\square}$

21 $\frac{12}{60} = \frac{1}{\square}$ **22** $\frac{15}{36} = \frac{5}{\square}$ **23** $\frac{16}{52} = \frac{4}{\square}$ **24** $\frac{22}{66} = \frac{1}{\square}$

25 $\frac{72}{80} = \frac{9}{\square}$ **26** $\frac{10}{35} = \frac{2}{\square}$ **27** $\frac{20}{70} = \frac{2}{\square}$ **28** $\frac{30}{65} = \frac{6}{\square}$

29 $\frac{15}{65} = \frac{3}{\square}$ **30** $\frac{24}{64} = \frac{3}{\square}$ **31** $\frac{27}{63} = \frac{3}{\square}$ **32** $\frac{28}{42} = \frac{2}{\square}$

33 $\frac{48}{64} = \frac{3}{\square}$ **34** $\frac{52}{65} = \frac{4}{\square}$ **35** $\frac{38}{57} = \frac{2}{\square}$ **36** $\frac{20}{80} = \frac{1}{\square}$

37 $\frac{32}{72} = \frac{4}{\square}$ **38** $\frac{54}{63} = \frac{6}{\square}$ **39** $\frac{66}{99} = \frac{2}{\square}$ **40** $\frac{84}{105} = \frac{4}{\square}$

UNIT 14 Equivalent Fractions

Equivalent fractions are formed when we mulitply or divide the numerator and denominator by the same number.

Equivalent fractions have the same value although they look different.

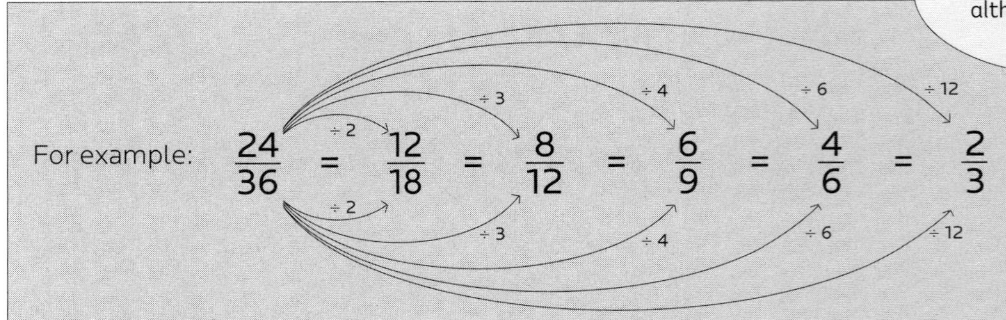

Complete the following equivalent fractions.

1 $\frac{20}{24} = \frac{\square}{6}$	**2** $\frac{16}{36} = \frac{\square}{9}$	**3** $\frac{18}{24} = \frac{\square}{4}$	**4** $\frac{20}{28} = \frac{\square}{7}$
5 $\frac{6}{18} = \frac{\square}{3}$	**6** $\frac{5}{35} = \frac{\square}{7}$	**7** $\frac{14}{21} = \frac{\square}{3}$	**8** $\frac{9}{36} = \frac{\square}{4}$
9 $\frac{8}{16} = \frac{\square}{2}$	**10** $\frac{8}{26} = \frac{\square}{13}$	**11** $\frac{10}{30} = \frac{\square}{3}$	**12** $\frac{20}{45} = \frac{\square}{9}$
13 $\frac{25}{45} = \frac{\square}{9}$	**14** $\frac{32}{60} = \frac{\square}{15}$	**15** $\frac{28}{42} = \frac{\square}{3}$	**16** $\frac{36}{63} = \frac{\square}{7}$
17 $\frac{80}{90} = \frac{\square}{9}$	**18** $\frac{63}{72} = \frac{\square}{8}$	**19** $\frac{64}{72} = \frac{\square}{9}$	**20** $\frac{52}{65} = \frac{\square}{5}$
21 $\frac{99}{110} = \frac{\square}{10}$	**22** $\frac{66}{99} = \frac{\square}{3}$	**23** $\frac{18}{54} = \frac{\square}{3}$	**24** $\frac{25}{35} = \frac{\square}{7}$
25 $\frac{30}{40} = \frac{\square}{4}$	**26** $\frac{60}{96} = \frac{\square}{8}$	**27** $\frac{28}{36} = \frac{\square}{9}$	**28** $\frac{33}{55} = \frac{\square}{5}$
29 $\frac{42}{56} = \frac{\square}{4}$	**30** $\frac{24}{28} = \frac{\square}{7}$	**31** $\frac{26}{39} = \frac{\square}{3}$	**32** $\frac{24}{40} = \frac{\square}{5}$
33 $\frac{24}{60} = \frac{\square}{5}$	**34** $\frac{32}{48} = \frac{\square}{3}$	**35** $\frac{36}{81} = \frac{\square}{9}$	**36** $\frac{18}{46} = \frac{\square}{23}$
37 $\frac{12}{40} = \frac{\square}{10}$	**38** $\frac{8}{28} = \frac{\square}{7}$	**39** $\frac{10}{25} = \frac{\square}{5}$	**40** $\frac{28}{80} = \frac{\square}{20}$
41 $\frac{22}{77} = \frac{\square}{7}$	**42** $\frac{84}{96} = \frac{\square}{8}$	**43** $\frac{34}{36} = \frac{\square}{18}$	**44** $\frac{38}{50} = \frac{\square}{25}$

UNIT 15 Comparing Fractions

It is very easy to compare fractions when they have the same denominator. The numerator then shows which fraction is smaller or larger.

The symbol $>$ means 'is greater than'.
The symbol $<$ means 'is less than'.

Example: Which is smaller?

(a) $\frac{3}{4}$ or $\frac{1}{4}$

(b) $\frac{6}{7}$ or $\frac{2}{7}$

Solution:

(a) Both the fractions have the same denominator and the numerator 1 is less than 4.

$\therefore \frac{1}{4} < \frac{3}{4}$

(b) Both the fractions have the same denominator and the numerator 2 is less than 6.

$\therefore \frac{2}{7} < \frac{6}{7}$

Use the correct sign $>$ or $<$ in between the fractions.

1 $\frac{2}{5}$ ☐ $\frac{4}{5}$	**2** $\frac{3}{6}$ ☐ $\frac{1}{6}$	**3** $\frac{3}{4}$ ☐ $\frac{1}{4}$	**4** $\frac{4}{9}$ ☐ $\frac{7}{9}$
5 $\frac{3}{8}$ ☐ $\frac{2}{8}$	**6** $\frac{5}{9}$ ☐ $\frac{3}{9}$	**7** $\frac{2}{11}$ ☐ $\frac{8}{11}$	**8** $\frac{9}{10}$ ☐ $\frac{3}{10}$
9 $\frac{5}{8}$ ☐ $\frac{7}{8}$	**10** $\frac{9}{13}$ ☐ $\frac{7}{13}$	**11** $\frac{6}{14}$ ☐ $\frac{5}{14}$	**12** $\frac{8}{17}$ ☐ $\frac{9}{17}$
13 $\frac{3}{10}$ ☐ $\frac{9}{10}$	**14** $\frac{6}{8}$ ☐ $\frac{3}{8}$	**15** $\frac{4}{15}$ ☐ $\frac{7}{15}$	**16** $\frac{6}{19}$ ☐ $\frac{5}{19}$
17 $\frac{3}{20}$ ☐ $\frac{2}{20}$	**18** $\frac{10}{21}$ ☐ $\frac{9}{21}$	**19** $\frac{11}{27}$ ☐ $\frac{13}{27}$	**20** $\frac{6}{35}$ ☐ $\frac{5}{35}$

Arrange each set of fractions in ascending order (from the smallest to the largest).

21 $\frac{3}{4}, \frac{1}{4}, \frac{2}{4}$ ________	**22** $\frac{4}{5}, \frac{1}{5}, \frac{3}{5}$ ________	**23** $\frac{3}{7}, \frac{1}{7}, \frac{2}{7}$ ________
24 $\frac{8}{9}, \frac{6}{9}, \frac{2}{9}$ ________	**25** $\frac{7}{11}, \frac{2}{11}, \frac{3}{11}$ ________	**26** $\frac{5}{13}, \frac{2}{13}, \frac{7}{13}$ ________
27 $\frac{8}{21}, \frac{3}{21}, \frac{2}{21}$ ________	**28** $\frac{6}{25}, \frac{11}{25}, \frac{3}{25}$ ________	**29** $\frac{8}{17}, \frac{1}{17}, \frac{2}{17}$ ________
30 $\frac{6}{15}, \frac{2}{15}, \frac{5}{15}$ ________	**31** $\frac{8}{14}, \frac{3}{14}, \frac{1}{14}$ ________	**32** $\frac{9}{13}, \frac{3}{13}, \frac{1}{13}$ ________
33 $\frac{9}{26}, \frac{5}{26}, \frac{1}{26}$ ________	**34** $\frac{2}{29}, \frac{3}{29}, \frac{1}{29}$ ________	**35** $\frac{7}{31}, \frac{2}{31}, \frac{5}{31}$ ________

UNIT 16 Comparing Fractions

To find which fraction is smaller or larger, make the denominators the same and then compare the numerators.

Example: Which is larger?

(a) $\frac{4}{5}$ or $\frac{2}{15}$ (b) $\frac{1}{6}$ or $\frac{2}{3}$ (c) $\frac{3}{14}$ or $\frac{1}{7}$

Solution:

(a) $\frac{4 \times 3}{5 \times 3}$ or $\frac{12}{15}$

$\frac{12}{15}$ or $\frac{2}{15}$

$\therefore \frac{12}{15} > \frac{2}{15}$

$\therefore \frac{4}{5} > \frac{2}{15}$

(b) $\frac{1}{6}$ or $\frac{2 \times 2}{3 \times 2}$

$\frac{1}{6}$ or $\frac{4}{6}$

$\therefore \frac{4}{6} > \frac{1}{6}$

$\therefore \frac{2}{3} > \frac{1}{6}$

(c) $\frac{3}{14}$ or $\frac{1}{7}$

$\frac{3}{14}$ or $\frac{1 \times 2}{7 \times 2}$

$\therefore \frac{3}{14}$ or $\frac{2}{14}$

$\therefore \frac{3}{14} > \frac{1}{7}$

To compare fractions, make the denominators the same.

Use the correct sign > or < in between the fractions.

1 $\frac{1}{2}$ ☐ $\frac{3}{4}$

2 $\frac{5}{6}$ ☐ $\frac{2}{3}$

3 $\frac{7}{8}$ ☐ $\frac{1}{4}$

4 $\frac{4}{9}$ ☐ $\frac{2}{3}$

5 $\frac{6}{15}$ ☐ $\frac{3}{5}$

6 $\frac{2}{9}$ ☐ $\frac{1}{3}$

7 $\frac{3}{20}$ ☐ $\frac{7}{10}$

8 $\frac{7}{10}$ ☐ $\frac{2}{5}$

9 $\frac{3}{14}$ ☐ $\frac{1}{2}$

10 $\frac{3}{8}$ ☐ $\frac{3}{4}$

11 $\frac{2}{15}$ ☐ $\frac{1}{3}$

12 $\frac{5}{12}$ ☐ $\frac{3}{4}$

13 $\frac{2}{9}$ ☐ $\frac{2}{3}$

14 $\frac{4}{9}$ ☐ $\frac{7}{18}$

15 $\frac{2}{13}$ ☐ $\frac{1}{26}$

16 $\frac{5}{7}$ ☐ $\frac{2}{14}$

17 $\frac{6}{11}$ ☐ $\frac{5}{22}$

18 $\frac{3}{10}$ ☐ $\frac{2}{5}$

UNIT 17 Comparing Fractions

To compare fractions with different denominators, we must change them to the same denominator; this is called the Lowest Common Denominator or Lowest Common Multiple (L.C.M.).

The Lowest Common Denominator of two or more numbers is the smallest common number that can be divided by all the denominators. For example:

- the Lowest Common Multiple of 2 and 3 is 6.
- the Lowest Common Multiple of 3, 4 and 8 is 24.
- the Lowest Common Multiple of 3, 4 and 5 is 60.

Example: Which is smaller?

(a) $\frac{1}{2}$ or $\frac{1}{3}$ (b) $\frac{2}{15}$ or $\frac{3}{10}$ (c) $\frac{4}{9}$ or $\frac{2}{5}$

Solution:

(a) $\frac{1 \times 3}{2 \times 3}$ or $\frac{1 \times 2}{3 \times 2}$

$\frac{3}{6}$ or $\frac{2}{6}$

$\therefore \frac{2}{6} < \frac{3}{6}$

$\therefore \frac{1}{3} < \frac{1}{2}$

(b) $\frac{2 \times 2}{15 \times 2}$ or $\frac{3 \times 3}{10 \times 3}$

$\frac{4}{30}$ or $\frac{9}{30}$

$\therefore \frac{4}{30} < \frac{9}{30}$

$\therefore \frac{2}{15} < \frac{3}{10}$

(c) $\frac{4 \times 5}{9 \times 5}$ or $\frac{2 \times 9}{5 \times 9}$

$\frac{20}{45}$ or $\frac{18}{45}$

$\therefore \frac{18}{45} < \frac{20}{45}$

$\therefore \frac{2}{5} < \frac{4}{9}$

Use the correct sign > or < in between the fractions.

1 $\frac{1}{3}$ ☐ $\frac{2}{5}$ **2** $\frac{3}{4}$ ☐ $\frac{2}{7}$ **3** $\frac{2}{3}$ ☐ $\frac{5}{8}$

4 $\frac{5}{6}$ ☐ $\frac{3}{5}$ **5** $\frac{3}{10}$ ☐ $\frac{2}{3}$ **6** $\frac{8}{12}$ ☐ $\frac{2}{15}$

7 $\frac{3}{8}$ ☐ $\frac{2}{5}$ **8** $\frac{4}{7}$ ☐ $\frac{3}{5}$ **9** $\frac{1}{2}$ ☐ $\frac{2}{3}$

10 $\frac{3}{4}$ ☐ $\frac{4}{5}$ **11** $\frac{7}{8}$ ☐ $\frac{8}{9}$ **12** $\frac{3}{5}$ ☐ $\frac{8}{9}$

13 $\frac{2}{15}$ ☐ $\frac{3}{4}$ **14** $\frac{1}{5}$ ☐ $\frac{3}{11}$ **15** $\frac{5}{6}$ ☐ $\frac{3}{10}$

UNIT 18 Comparing Fractions

To compare fractions when they have different denominators, we should change them into equivalent fractions with the same denominator.

Once the denominators are the same, the numerator shows which fraction is smaller or larger.

Example: Which fraction is larger, $\frac{2}{3}$ or $\frac{5}{6}$?

Solution: Since the denominators of the fraction are different, we should change them into equivalent fractions with the same denominator. $\frac{2 \times 2}{3 \times 2} = \frac{4}{6}$

Now $\frac{4}{6}$ and $\frac{5}{6}$ have the same denominator and the numerator 5 is greater than 4.

$\therefore \ \frac{5}{6} > \frac{4}{6}$

$\therefore \ \frac{5}{6} > \frac{2}{3}$

If we multiply or divide the numerator and denominator by the same number, equivalent fractions are formed.

Use >, < or = to show comparisons between the following pairs of fractions.

1 $\frac{1}{2}$ ☐ $\frac{1}{3}$ **2** $\frac{1}{3}$ ☐ $\frac{1}{4}$ **3** $\frac{2}{3}$ ☐ $\frac{3}{4}$ **4** $\frac{2}{3}$ ☐ $\frac{4}{5}$

5 $\frac{5}{6}$ ☐ $\frac{5}{8}$ **6** $\frac{3}{6}$ ☐ $\frac{1}{2}$ **7** $\frac{4}{5}$ ☐ $\frac{3}{15}$ **8** $\frac{6}{8}$ ☐ $\frac{3}{4}$

9 $\frac{13}{15}$ ☐ $\frac{3}{5}$ **10** $\frac{9}{10}$ ☐ $\frac{3}{5}$ **11** $\frac{2}{3}$ ☐ $\frac{6}{9}$ **12** $\frac{12}{15}$ ☐ $\frac{4}{5}$

13 $\frac{25}{100}$ ☐ $\frac{1}{5}$ **14** $\frac{6}{8}$ ☐ $\frac{2}{3}$ **15** $\frac{6}{7}$ ☐ $\frac{5}{14}$ **16** $\frac{9}{10}$ ☐ $\frac{3}{5}$

17 $\frac{3}{6}$ ☐ $\frac{2}{3}$ **18** $\frac{5}{6}$ ☐ $\frac{3}{4}$ **19** $\frac{3}{5}$ ☐ $\frac{7}{10}$ **20** $\frac{4}{6}$ ☐ $\frac{3}{5}$

Arrange each set of fractions in ascending order (from the smallest to the largest).

21 $\frac{1}{2}, \frac{2}{3}, \frac{1}{3}$ ____________ **22** $\frac{4}{5}, \frac{6}{7}, \frac{5}{6}$ ____________ **23** $\frac{3}{3}, \frac{3}{4}, \frac{3}{5}$ ____________

24 $\frac{7}{10}, \frac{8}{20}, \frac{80}{100}$ ____________ **25** $\frac{1}{5}, \frac{1}{3}, \frac{1}{8}$ ____________ **26** $\frac{2}{4}, \frac{3}{8}, \frac{3}{4}$ ____________

27 $\frac{2}{4}, \frac{3}{5}, \frac{3}{10}$ ____________ **28** $\frac{1}{3}, \frac{1}{6}, \frac{1}{9}$ ____________ **29** $\frac{4}{5}, \frac{7}{20}, \frac{9}{10}$ ____________

30 $\frac{3}{8}, \frac{2}{5}, \frac{4}{7}$ ____________ **31** $\frac{1}{4}, \frac{2}{5}, \frac{3}{7}$ ____________ **32** $\frac{3}{4}, \frac{3}{5}, \frac{4}{9}$ ____________

Quick Questions

This unit of 'Quick Questions' is designed to make sure that you think fast and remain in touch with the previous work of the section.

If two fractions have the same denominator then the numerator shows which fraction is smaller or larger.

Complete the following equivalent fractions.

1 $\frac{1}{2} = \frac{2}{\square}$ **2** $\frac{1}{3} = \frac{2}{\square}$ **3** $\frac{1}{4} = \frac{2}{\square}$ **4** $\frac{1}{5} = \frac{2}{\square}$

5 $\frac{1}{3} = \frac{3}{\square}$ **6** $\frac{1}{5} = \frac{3}{\square}$ **7** $\frac{1}{7} = \frac{3}{\square}$ **8** $\frac{1}{9} = \frac{3}{\square}$

9 $\frac{1}{2} = \frac{\square}{6}$ **10** $\frac{1}{4} = \frac{\square}{8}$ **11** $\frac{1}{6} = \frac{\square}{12}$ **12** $\frac{1}{8} = \frac{\square}{32}$

13 $\frac{2}{3} = \frac{\square}{6}$ **14** $\frac{3}{4} = \frac{\square}{12}$ **15** $\frac{4}{5} = \frac{\square}{15}$ **16** $\frac{5}{6} = \frac{\square}{24}$

17 $\frac{4}{6} = \frac{2}{\square}$ **18** $\frac{6}{8} = \frac{3}{\square}$ **19** $\frac{8}{10} = \frac{4}{\square}$ **20** $\frac{10}{12} = \frac{5}{\square}$

21 $\frac{3}{6} = \frac{1}{\square}$ **22** $\frac{6}{9} = \frac{2}{\square}$ **23** $\frac{4}{12} = \frac{1}{\square}$ **24** $\frac{8}{24} = \frac{1}{\square}$

25 $\frac{8}{20} = \frac{\square}{5}$ **26** $\frac{10}{30} = \frac{\square}{3}$ **27** $\frac{5}{15} = \frac{\square}{3}$ **28** $\frac{7}{21} = \frac{\square}{3}$

29 $\frac{14}{21} = \frac{\square}{3}$ **30** $\frac{12}{16} = \frac{\square}{4}$ **31** $\frac{6}{10} = \frac{\square}{5}$ **32** $\frac{10}{14} = \frac{\square}{7}$

Arrange each set of fractions in ascending order (from the smallest to the largest).

33 $\frac{3}{6}, \frac{1}{6}, \frac{6}{6}$ ____________ **34** $\frac{3}{5}, \frac{1}{5}, \frac{2}{5}$ ____________ **35** $\frac{4}{7}, \frac{1}{7}, \frac{6}{7}$ ____________

36 $\frac{2}{9}, \frac{1}{9}, \frac{5}{9}$ ____________ **37** $\frac{8}{11}, \frac{6}{11}, \frac{5}{11}$ ____________ **38** $\frac{3}{3}, \frac{1}{3}, \frac{2}{3}$ ____________

39 $\frac{7}{12}, \frac{8}{12}, \frac{3}{12}$ ____________ **40** $\frac{5}{13}, \frac{2}{13}, \frac{4}{13}$ ____________ **41** $\frac{6}{9}, \frac{5}{9}, \frac{3}{9}$ ____________

42 $\frac{2}{14}, \frac{9}{14}, \frac{8}{14}$ ____________ **43** $\frac{3}{15}, \frac{1}{15}, \frac{2}{15}$ ____________ **44** $\frac{7}{16}, \frac{8}{16}, \frac{5}{16}$ ____________

45 $\frac{3}{4}, \frac{1}{4}, \frac{4}{4}$ ____________ **46** $\frac{5}{8}, \frac{3}{8}, \frac{4}{8}$ ____________ **47** $\frac{9}{10}, \frac{7}{10}, \frac{10}{10}$ ____________

UNIT 20 Stay in Touch

The unit 'Stay In Touch' is intended to keep you in touch with what you have done before.

It also ensures that all types of questions in this section receive constant revision.

Complete the following equivalent fractions.

1 $\frac{5}{7} = \frac{15}{\square}$ 2 $\frac{3}{5} = \frac{12}{\square}$ 3 $\frac{7}{9} = \frac{28}{\square}$ 4 $\frac{9}{11} = \frac{45}{\square}$

5 $\frac{2}{12} = \frac{12}{\square}$ 6 $\frac{3}{15} = \frac{21}{\square}$ 7 $\frac{4}{14} = \frac{24}{\square}$ 8 $\frac{6}{16} = \frac{18}{\square}$

9 $\frac{8}{14} = \frac{\square}{28}$ 10 $\frac{9}{15} = \frac{\square}{45}$ 11 $\frac{12}{18} = \frac{\square}{36}$ 12 $\frac{15}{21} = \frac{\square}{42}$

13 $\frac{18}{24} = \frac{\square}{72}$ 14 $\frac{21}{27} = \frac{\square}{81}$ 15 $\frac{24}{30} = \frac{\square}{90}$ 16 $\frac{30}{42} = \frac{\square}{84}$

17 $\frac{8}{18} = \frac{4}{\square}$ 18 $\frac{14}{24} = \frac{7}{\square}$ 19 $\frac{20}{30} = \frac{2}{\square}$ 20 $\frac{26}{40} = \frac{13}{\square}$

21 $\frac{34}{50} = \frac{17}{\square}$ 22 $\frac{42}{60} = \frac{7}{\square}$ 23 $\frac{50}{70} = \frac{5}{\square}$ 24 $\frac{58}{82} = \frac{29}{\square}$

25 $\frac{24}{36} = \frac{\square}{3}$ 26 $\frac{22}{30} = \frac{\square}{15}$ 27 $\frac{20}{24} = \frac{\square}{6}$ 28 $\frac{30}{34} = \frac{\square}{17}$

29 $\frac{40}{44} = \frac{\square}{11}$ 30 $\frac{50}{54} = \frac{\square}{27}$ 31 $\frac{64}{72} = \frac{\square}{9}$ 32 $\frac{75}{95} = \frac{\square}{19}$

Arrange each set of fractions in descending order (from the largest to the smallest).

33 $\frac{2}{7}, \frac{1}{7}, \frac{5}{7}$ ______ 34 $\frac{8}{12}, \frac{7}{12}, \frac{3}{4}$ ______ 35 $\frac{7}{21}, \frac{3}{21}, \frac{9}{21}$ ______

36 $\frac{2}{5}, \frac{10}{15}, \frac{8}{15}$ ______ 37 $\frac{2}{3}, \frac{1}{4}, \frac{3}{8}$ ______ 38 $\frac{2}{3}, \frac{3}{4}, \frac{6}{12}$ ______

39 $\frac{1}{2}, \frac{3}{4}, \frac{5}{8}$ ______ 40 $\frac{8}{5}, \frac{6}{10}, \frac{15}{20}$ ______ 41 $\frac{8}{24}, \frac{1}{4}, \frac{5}{8}$ ______

42 $\frac{1}{6}, \frac{17}{30}, \frac{2}{5}$ ______ 43 $\frac{4}{5}, \frac{3}{10}, \frac{8}{15}$ ______ 44 $\frac{2}{5}, \frac{3}{15}, \frac{8}{10}$ ______

45 $\frac{8}{9}, \frac{1}{3}, \frac{7}{18}$ ______ 46 $\frac{2}{7}, \frac{5}{14}, \frac{8}{21}$ ______ 47 $\frac{3}{8}, \frac{3}{4}, \frac{7}{8}$ ______

Stop. Revise. Check.

- A fraction is also called a rational number.
- Equivalent fractions have the same value and represent the same number but have different numerators and denominators. e.g. $\frac{1}{2} = \frac{2}{4} = \frac{8}{16}$
- In order to obtain equivalent fractions, multiply or divide the numerator and denominator by the same number. e.g. (a)

$$\frac{1}{2} = \frac{2}{4} = \frac{8}{16}$$

($\times 2$ from $\frac{1}{2}$ to $\frac{2}{4}$; $\times 8$ from $\frac{1}{2}$ to $\frac{8}{16}$)

(b)

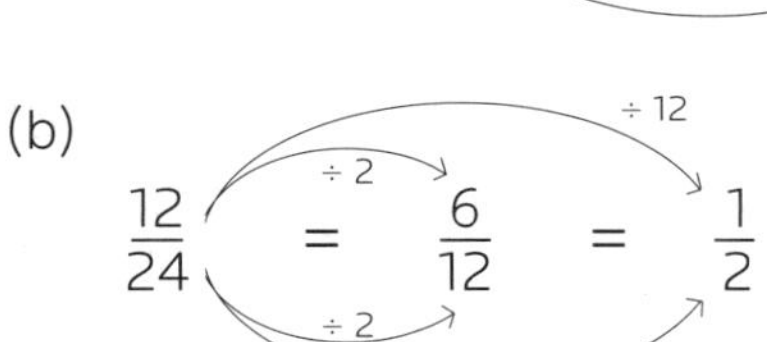

- The symbol $>$ means 'is greater than'. e.g. $\frac{3}{4} > \frac{1}{2}$
- The symbol $<$ means 'is less than'. e.g. $\frac{1}{4} < 1\frac{1}{2}$
- The Lowest Common Denominator (also called the Lowest Common Multiple) of two or more numbers is the smallest number that can be divided by all the denominators.
  e.g. the Lowest Common Multiple of 2, 3 and 4 is 12.
- When fractions have the same denominators, the numerator shows which fraction is smaller or larger.
  e.g. (a) $\frac{1}{5} < \frac{3}{5}$ (b) $\frac{2}{7} < \frac{6}{7}$ (c) $\frac{5}{9} > \frac{2}{9}$
- When fractions have different denominators, change these into equivalent fractions with the same denominator. The numerator then shows which fraction is smaller or larger.

Example: Which fraction is larger:

(a) $\frac{2}{3}$ or $\frac{7}{9}$? (b) $\frac{4}{9}$ or $\frac{2}{5}$?

Solution: (a) $\frac{2}{3} = \frac{2 \times 3}{3 \times 3} = \frac{6}{9}$

$\therefore \frac{7}{9} > \frac{6}{9}$

$\frac{7}{9} > \frac{2}{3}$

(b) $\frac{4 \times 5}{9 \times 5}$ or $\frac{2 \times 9}{5 \times 9}$

$\frac{20}{45}$ or $\frac{18}{45}$

$\frac{20}{45} > \frac{18}{45}$

$\frac{4}{9} > \frac{2}{5}$

UNIT 22 Test on Section 2

Instructions: Do not use a calculator.
Attempt all questions.

Time allowed: 25 minutes **Total marks:** 54

Marks

Question 1: Complete the following equivalent fractions.

a $\frac{6}{7} = \frac{18}{\square}$ **b** $\frac{5}{9} = \frac{20}{\square}$ **c** $\frac{6}{14} = \frac{18}{\square}$ **d** $\frac{8}{15} = \frac{40}{\square}$ 4

e $\frac{9}{12} = \frac{27}{\square}$ **f** $\frac{15}{18} = \frac{60}{\square}$ **g** $\frac{21}{28} = \frac{42}{\square}$ **h** $\frac{35}{40} = \frac{70}{\square}$ 4

Question 2: Find the missing numbers.

a $\frac{2}{5} = \frac{\square}{20}$ **b** $\frac{3}{4} = \frac{\square}{24}$ **c** $\frac{5}{6} = \frac{\square}{48}$ **d** $\frac{7}{8} = \frac{\square}{72}$ 4

e $\frac{9}{12} = \frac{\square}{36}$ **f** $\frac{10}{12} = \frac{\square}{60}$ **g** $\frac{15}{18} = \frac{\square}{54}$ **h** $\frac{12}{13} = \frac{\square}{52}$ 4

Question 3: Complete the following equivalent fractions.

a $\frac{16}{24} = \frac{2}{\square}$ **b** $\frac{45}{60} = \frac{3}{\square}$ **c** $\frac{20}{45} = \frac{\square}{9}$ **d** $\frac{60}{72} = \frac{\square}{6}$ 4

e $\frac{18}{27} = \frac{2}{\square}$ **f** $\frac{36}{54} = \frac{2}{\square}$ **g** $\frac{33}{55} = \frac{\square}{5}$ **h** $\frac{64}{88} = \frac{\square}{11}$ 4

Question 4: Arrange each set of fractions in descending order.

a $\frac{8}{11}, \frac{2}{11}, \frac{5}{11}$ ________ **b** $\frac{2}{21}, \frac{5}{21}, \frac{3}{21}$ ________ **c** $\frac{8}{37}, \frac{3}{37}, \frac{5}{37}$ ________ 6

d $\frac{2}{3}, \frac{4}{12}, \frac{5}{6}$ ________ **e** $\frac{2}{5}, \frac{1}{15}, \frac{7}{15}$ ________ **f** $\frac{3}{8}, \frac{3}{4}, \frac{7}{16}$ ________ 6

Question 5: Arrange each set of fractions in ascending order.

a $\frac{1}{8}, \frac{3}{8}, \frac{2}{8}$ ________ **b** $\frac{5}{11}, \frac{3}{11}, \frac{9}{11}$ ________ **c** $\frac{6}{15}, \frac{3}{15}, \frac{7}{15}$ ________ 6

d $\frac{1}{2}, \frac{3}{4}, \frac{7}{8}$ ________ **e** $\frac{5}{6}, \frac{7}{12}, \frac{2}{3}$ ________ **f** $\frac{1}{3}, \frac{5}{9}, \frac{5}{6}$ ________ 6

g $\frac{3}{4}, \frac{5}{8}, \frac{11}{12}$ ________ **h** $\frac{1}{2}, \frac{3}{5}, \frac{3}{10}$ ________ **i** $\frac{3}{4}, \frac{5}{6}, \frac{3}{8}$ ________ 6

Total marks = ___ / 54

SECTION 3

UNIT 23 Addition of Fractions (Fractions with the Same Denominator)

It is very easy to add fractions with the same denominator. We simply add the numerators then, if it is possible, we also simplify the answer.

Example: Add the following fractions.

(a) $\frac{3}{7} + \frac{1}{7}$ (b) $\frac{1}{4} + \frac{1}{4}$

Solution:

(a) $\frac{3}{7} + \frac{1}{7} = \frac{3+1}{7} = \frac{4}{7}$ (b) $\frac{1}{4} + \frac{1}{4} = \frac{1+1}{4} = \frac{2}{4} = \frac{2 \div 2}{4 \div 2} = \frac{1}{2}$

Add the following fractions.

1 $\frac{1}{3} + \frac{1}{3} = \frac{\quad}{3}$

2 $\frac{2}{5} + \frac{1}{5} = \frac{\quad}{5}$

3 $\frac{4}{7} + \frac{2}{7} = \frac{\quad}{7}$

4 $\frac{3}{9} + \frac{4}{9} = \frac{\quad}{9}$

5 $\frac{4}{11} + \frac{6}{11} = \frac{\quad}{11}$

6 $\frac{3}{10} + \frac{4}{10} = \frac{\quad}{10}$

7 $\frac{1}{4} + \frac{2}{4} = \frac{\quad}{4}$

8 $\frac{1}{8} + \frac{4}{8} = \frac{\quad}{8}$

9 $\frac{4}{15} + \frac{7}{15} = \frac{\quad}{15}$

10 $\frac{5}{12} + \frac{2}{12} = \frac{\quad}{12}$

11 $\frac{6}{17} + \frac{3}{17} = \frac{\quad}{17}$

12 $\frac{4}{21} + \frac{7}{21} = \frac{\quad}{21}$

13 $\frac{2}{9} + \frac{1}{9} = \frac{\quad}{9} = \underline{\qquad\qquad}$

14 $\frac{3}{8} + \frac{1}{8} = \frac{\quad}{8} = \underline{\qquad\qquad}$

15 $\frac{5}{12} + \frac{3}{12} = \frac{\quad}{12} = \underline{\qquad\qquad}$

16 $\frac{7}{15} + \frac{2}{15} = \frac{\quad}{15} = \underline{\qquad\qquad}$

17 $\frac{9}{28} + \frac{5}{28} = \frac{\quad}{28} = \underline{\qquad\qquad}$

18 $\frac{5}{30} + \frac{13}{30} = \frac{\quad}{30} = \underline{\qquad\qquad}$

19 $\frac{2}{27} + \frac{7}{27} = \frac{\quad}{27} = \underline{\qquad\qquad}$

20 $\frac{3}{20} + \frac{9}{20} = \frac{\quad}{20} = \underline{\qquad\qquad}$

21 $\frac{4}{25} + \frac{16}{25} = \frac{\quad}{25} = \underline{\qquad\qquad}$

22 $\frac{9}{40} + \frac{21}{40} = \frac{\quad}{40} = \underline{\qquad\qquad}$

23 $\frac{7}{24} + \frac{11}{24} = \frac{\quad}{24} = \underline{\qquad\qquad}$

24 $\frac{5}{18} + \frac{7}{18} = \frac{\quad}{18} = \underline{\qquad\qquad}$

UNIT 24 Addition of Fractions (Fractions with Different Denominators)

In order to add fractions with different denominators, change them to equivalent fractions with the same denominator. Then simply add the numerators and, if possible, simplify the result.

Always reduce your answer to the lowest terms.

Example: Add the following fractions.

(a) $\frac{5}{6} + \frac{1}{12}$ (b) $\frac{2}{3} + \frac{4}{5}$

Solution:

(a) $\frac{5}{6} + \frac{1}{12} = \frac{5 \times 2}{6 \times 2} + \frac{1}{12} = \frac{10}{12} + \frac{1}{12} = \frac{10+1}{12} = \frac{11}{12}$

(b) $\frac{2}{3} + \frac{4}{5} = \frac{2 \times 5}{3 \times 5} + \frac{4 \times 3}{5 \times 3} = \frac{10}{15} + \frac{12}{15} = \frac{10+12}{15} = \frac{22}{15} = 1\frac{7}{15}$

Add the following fractions.

1 $\frac{3}{5} + \frac{3}{10} =$ ______

2 $\frac{3}{8} + \frac{1}{2} =$ ______

3 $\frac{1}{2} + \frac{4}{5} =$ ______

4 $\frac{1}{3} + \frac{1}{2} =$ ______

5 $\frac{1}{3} + \frac{1}{4} =$ ______

6 $\frac{1}{7} + \frac{3}{14} =$ ______

7 $\frac{3}{20} + \frac{2}{5} =$ ______

8 $\frac{3}{5} + \frac{2}{15} =$ ______

9 $\frac{2}{9} + \frac{2}{3} =$ ______

10 $\frac{2}{11} + \frac{9}{22} =$ ______

11 $\frac{1}{2} + \frac{5}{24} =$ ______

12 $\frac{1}{3} + \frac{5}{18} =$ ______

13 $\frac{2}{6} + \frac{1}{3} =$ ______

14 $\frac{2}{8} + \frac{1}{2} =$ ______

15 $\frac{2}{4} + \frac{2}{6} =$ ______

16 $\frac{5}{8} + \frac{1}{4} =$ ______

17 $\frac{3}{27} + \frac{7}{9} =$ ______

18 $\frac{2}{8} + \frac{2}{6} =$ ______

19 $\frac{18}{27} + \frac{2}{3} =$ ______

20 $\frac{4}{15} + \frac{2}{5} =$ ______

21 $\frac{7}{20} + \frac{2}{5} =$ ______

UNIT 25 Addition of Fractions

In order to add three or more fractions with the same denominators, simply add the numerators and, if possible, simplify.

To add fractions with different denominators, change them to equivalent fractions with the same denominator, then add the numerators and simplify.

Example: Add the following fractions.

(a) $\frac{2}{5} + \frac{1}{5} + \frac{1}{5}$ (b) $\frac{2}{7} + \frac{1}{7} + \frac{3}{14}$

Solution:

(a) $\frac{2}{5} + \frac{1}{5} + \frac{1}{5} = \frac{2+1+1}{5}$

$= \frac{4}{5}$

(b) $\frac{2}{7} + \frac{1}{7} + \frac{3}{14} = \frac{2 \times 2}{7 \times 2} + \frac{1 \times 2}{7 \times 2} + \frac{3}{14}$

$= \frac{4}{14} + \frac{2}{14} + \frac{3}{14}$

$= \frac{4+2+3}{14}$

$= \frac{9}{14}$

Add the following fractions.

1 $\frac{2}{3} + \frac{2}{3} + \frac{1}{3} =$ ________ **2** $\frac{3}{20} + \frac{7}{20} + \frac{1}{20} =$ ________ **3** $\frac{1}{3} + \frac{2}{3} + \frac{1}{7} =$ ________

4 $\frac{7}{20} + \frac{1}{10} + \frac{3}{5} =$ ________ **5** $\frac{4}{15} + \frac{2}{5} + \frac{3}{5} =$ ________ **6** $\frac{13}{30} + \frac{1}{10} + \frac{2}{5} =$ ________

7 $\frac{3}{4} + \frac{1}{2} + \frac{3}{8} =$ ________ **8** $\frac{5}{6} + \frac{3}{4} + \frac{5}{8} =$ ________ **9** $\frac{2}{9} + \frac{3}{5} + \frac{7}{10} =$ ________

10 $\frac{1}{2} + \frac{2}{3} + \frac{3}{4} =$ ________ **11** $\frac{2}{5} + \frac{3}{4} + \frac{9}{10} =$ ________ **12** $\frac{1}{4} + \frac{5}{6} + \frac{7}{8} =$ ________

UNIT 26 Addition of Fractions

(Fractions with Mixed Numbers)

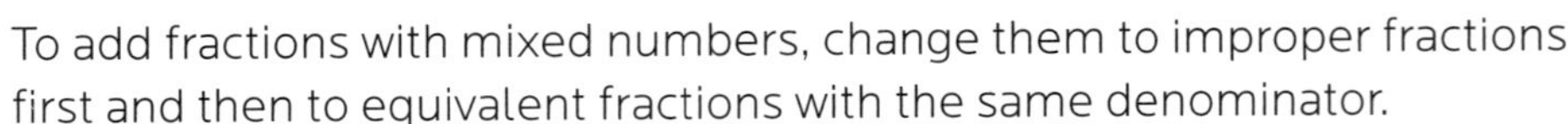

To add fractions with mixed numbers, change them to improper fractions first and then to equivalent fractions with the same denominator.

A mixed number has a whole number and a proper fraction part.

Example: Add the following fractions.

(a) $4\frac{3}{4} + 1\frac{1}{2} = \frac{19}{4} + \frac{3}{2} = \frac{19}{4} + \frac{3 \times 2}{2 \times 2}$

$= \frac{19}{4} + \frac{6}{4} = \frac{19 + 6}{4}$

$= \frac{25}{4} = 6\frac{1}{4}$

(b) $3\frac{1}{2} + 4\frac{1}{6} = \frac{7}{2} + \frac{25}{6} = \frac{7 \times 3}{2 \times 3} + \frac{25}{6}$

$= \frac{21}{6} + \frac{25}{6} = \frac{46}{6}$

$= \frac{23}{3} = 7\frac{2}{3}$

Add the following fractions.

1 $8\frac{1}{4} + 2\frac{1}{2} =$ ________

2 $6\frac{1}{4} + 3\frac{3}{4} =$ ________

3 $1 + \frac{5}{24} =$ ________

4 $5\frac{2}{3} + 3\frac{1}{4} =$ ________

5 $2 + \frac{1}{8} =$ ________

6 $6\frac{3}{4} + 3\frac{1}{2} =$ ________

7 $5\frac{2}{5} + 3 =$ ________

8 $2\frac{1}{2} + 1\frac{1}{2} =$ ________

9 $3\frac{1}{4} + 5\frac{1}{2} =$ ________

10 $2\frac{3}{5} + 4\frac{2}{5} =$ ________

11 $2\frac{1}{2} + 2\frac{1}{3} =$ ________

12 $3\frac{1}{4} + 2\frac{1}{2} =$ ________

13 $2\frac{1}{4} + 1\frac{4}{5} =$ ________

14 $5\frac{2}{5} + 1\frac{3}{4} =$ ________

15 $4\frac{1}{2} + 3\frac{5}{6} =$ ________

16 $2\frac{1}{9} + 1\frac{1}{3} =$ ________

17 $6\frac{1}{2} + 2\frac{1}{4} =$ ________

18 $8\frac{1}{3} + 2\frac{3}{4} =$ ________

UNIT 27 Subtraction of Fractions

(Fractions with the Same Denominator)

It is very easy to subtract fractions with the same denominator.
We simply subtract the numerators and, if possible, the result is simplified.

The top part of the fraction is called the numerator.

Example: Subtract the following fractions.

(a) $\frac{3}{11} - \frac{1}{11}$

(b) $\frac{5}{8} - \frac{3}{8}$

Solution:

(a) $\frac{3}{11} - \frac{1}{11} = \frac{3-1}{11}$

$= \frac{2}{11}$

(b) $\frac{5}{8} - \frac{3}{8} = \frac{5-3}{8} = \frac{2}{8}$

$= \frac{2 \div 2}{8 \div 2} = \frac{1}{4}$

Subtract the following fractions.

1 $\frac{2}{3} - \frac{1}{3}$ = ______

2 $\frac{3}{7} - \frac{1}{7}$ = ______

3 $\frac{5}{9} - \frac{3}{9}$ = ______

4 $\frac{6}{13} - \frac{4}{13}$ = ______

5 $\frac{9}{17} - \frac{6}{17}$ = ______

6 $\frac{8}{25} - \frac{5}{25}$ = ______

7 $\frac{17}{35} - \frac{9}{35}$ = ______

8 $\frac{6}{15} - \frac{4}{15}$ = ______

9 $\frac{15}{42} - \frac{10}{42}$ = ______

10 $\frac{12}{37} - \frac{7}{37}$ = ______

11 $\frac{8}{15} - \frac{3}{15}$ = ______

12 $\frac{19}{26} - \frac{6}{26}$ = ______

13 $\frac{7}{12} - \frac{3}{12}$ = ______

14 $\frac{5}{27} - \frac{2}{27}$ = ______

15 $\frac{11}{15} - \frac{6}{15}$ = ______

16 $\frac{20}{33} - \frac{9}{33}$ = ______

17 $\frac{18}{35} - \frac{11}{35}$ = ______

18 $\frac{9}{20} - \frac{4}{20}$ = ______

19 $\frac{6}{10} - \frac{2}{10}$ = ______

20 $\frac{13}{28} - \frac{9}{28}$ = ______

21 $\frac{5}{8} - \frac{2}{8} - \frac{1}{8}$ = ______

22 $\frac{7}{15} - \frac{3}{15} - \frac{1}{15}$ = ______

23 $\frac{11}{16} - \frac{3}{16} - \frac{2}{16}$ = ______

24 $\frac{7}{27} - \frac{3}{27} - \frac{1}{27}$ = ______

UNIT 28 Subtraction of Fractions

(Fractions with Different Denominators)

In order to subtract fractions with different denominators, change them to equivalent fractions with the same denominator.
Then simply subtract the numerators and, if possible, simplify the result.

Once the denominators are the same, just subtract the numerators.

Example: Subtract the following fractions.

(a) $\frac{5}{8} - \frac{1}{4}$ (b) $\frac{8}{10} - \frac{1}{3}$

Solution:

(a) $\frac{5}{8} - \frac{1}{4} = \frac{5}{8} - \frac{1 \times 2}{4 \times 2} = \frac{5}{8} - \frac{2}{8} = \frac{3}{8}$

(b) $\frac{8}{10} - \frac{1}{3} = \frac{8}{10} \times \frac{3}{3} - \frac{1}{3} \times \frac{10}{10} = \frac{24}{30} - \frac{10}{30} = \frac{14}{30} = \frac{7}{15}$

Subtract the following fractions.

1 $\frac{3}{5} - \frac{3}{10} =$ ________

2 $\frac{9}{10} - \frac{1}{2} =$ ________

3 $\frac{1}{4} - \frac{1}{5} =$ ________

4 $\frac{9}{10} - \frac{1}{5} =$ ________

5 $\frac{11}{20} - \frac{7}{20} =$ ________

6 $\frac{7}{8} - \frac{2}{3} =$ ________

7 $\frac{93}{100} - \frac{3}{4} =$ ________

8 $\frac{1}{2} - \frac{1}{10} =$ ________

9 $\frac{4}{5} - \frac{1}{3} =$ ________

10 $\frac{3}{4} - \frac{1}{2} =$ ________

11 $\frac{8}{10} - \frac{2}{5} =$ ________

12 $\frac{5}{6} - \frac{2}{3} =$ ________

13 $\frac{14}{16} - \frac{3}{8} =$ ________

14 $\frac{5}{7} - \frac{8}{14} =$ ________

15 $\frac{27}{20} - \frac{2}{5} =$ ________

16 $\frac{1}{3} - \frac{1}{15} =$ ________

17 $\frac{3}{4} - \frac{1}{6} =$ ________

18 $\frac{3}{7} - \frac{1}{9} =$ ________

19 $\frac{6}{7} - \frac{1}{3} =$ ________

20 $\frac{9}{14} - \frac{2}{7} =$ ________

21 $\frac{18}{25} - \frac{2}{5} + \frac{3}{5} =$ ________

UNIT 29 Subtraction of Fractions

(Fractions with Mixed Numbers)

To subtract fractions with mixed numbers, change them to improper fractions first and then to equivalent fractions with the same denominator.

Before subtracting, express each fraction with the same denominator.

Example: Subtract the following fractions.

(a) $8\frac{1}{4} - 2\frac{1}{2}$ (b) $5\frac{1}{2} - \frac{7}{8}$

Solution:

(a) $8\frac{1}{4} - 2\frac{1}{2} = \frac{33}{4} - \frac{5}{2} = \frac{33}{4} - \frac{10}{4}$

$= \frac{23}{4} = 5\frac{3}{4}$

(b) $5\frac{1}{2} - \frac{7}{8} = \frac{11}{2} - \frac{7}{8} = \frac{44}{8} - \frac{7}{8}$

$= \frac{37}{8} = 4\frac{5}{8}$

Subtract the following fractions.

1 $5\frac{2}{3} - 3\frac{1}{4} =$ __________

2 $8\frac{2}{5} - 4 =$ __________

3 $5\frac{7}{8} - 5\frac{3}{8} =$ __________

4 $12\frac{7}{10} - 5\frac{2}{3} =$ __________

5 $4\frac{1}{2} - 3\frac{1}{3} =$ __________

6 $8\frac{1}{4} - 6\frac{2}{3} =$ __________

7 $7\frac{5}{6} - 2\frac{3}{4} =$ __________

8 $9\frac{1}{3} - 3\frac{5}{9} =$ __________

9 $6\frac{2}{3} - 4\frac{1}{9} =$ __________

10 $8\frac{3}{4} - 5\frac{1}{2} =$ __________

11 $16\frac{2}{5} - 12\frac{3}{4} =$ __________

12 $7\frac{2}{9} - 5\frac{3}{4} =$ __________

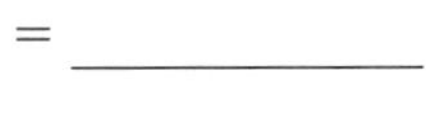

13 $9\frac{1}{2} - 6\frac{1}{3} =$ __________

14 $12\frac{5}{6} - 6\frac{3}{4} =$ __________

15 $2\frac{11}{12} - 2\frac{3}{8} =$ __________

UNIT 30 Quick Questions

This unit of 'Quick Questions' is designed to make sure that you think fast and remain in touch with the previous work of this section.

Simplify the following.

1 $\frac{1}{2} + \frac{1}{2} =$ ____________________

2 $\frac{1}{3} + \frac{1}{3} =$ ____________________

3 $\frac{3}{5} - \frac{1}{5} =$ ____________________

4 $\frac{5}{6} - \frac{1}{6} =$ ____________________

5 $\frac{1}{4} + \frac{1}{2} =$ ____________________

6 $\frac{1}{2} + \frac{2}{3} =$ ____________________

7 $\frac{1}{3} + \frac{2}{3} =$ ____________________

8 $\frac{1}{4} + \frac{1}{5} =$ ____________________

9 $\frac{1}{2} + \frac{1}{3} =$ ____________________

10 $\frac{3}{4} + \frac{1}{4} =$ ____________________

11 $\frac{5}{6} - \frac{3}{6} =$ ____________________

12 $\frac{1}{2} - \frac{1}{3} =$ ____________________

13 $\frac{5}{9} + \frac{2}{9} =$ ____________________

14 $\frac{1}{2} - \frac{1}{4} =$ ____________________

15 $\frac{1}{5} + \frac{3}{5} =$ ____________________

16 $\frac{1}{3} + \frac{1}{5} =$ ____________________

17 $\frac{1}{3} + \frac{1}{4} =$ ____________________

18 $\frac{1}{5} + \frac{2}{5} =$ ____________________

19 $\frac{3}{5} + \frac{3}{5} =$ ____________________

20 $\frac{1}{4} + \frac{2}{5} =$ ____________________

21 $\frac{1}{2} + \frac{5}{6} =$ ____________________

22 $\frac{1}{2} + \frac{3}{7} =$ ____________________

23 $\frac{8}{9} - \frac{2}{3} =$ ____________________

24 $\frac{7}{12} - \frac{1}{6} =$ ____________________

25 $\frac{3}{8} - \frac{1}{4} =$ ____________________

26 $\frac{7}{10} - \frac{2}{10} =$ ____________________

To add fractions with the same denominator, simply add the numerators.

UNIT 31 Stay in Touch

The unit 'Stay In Touch' is intended to keep you in touch with what you have done before. It also ensures that all types of questions in this section receive constant revision.

Simplify the following.

1 $\frac{7}{8}+\frac{1}{8}$ = ______

2 $\frac{7}{12}+\frac{2}{7}$ = ______

3 $\frac{5}{11}+\frac{3}{8}$ = ______

4 $\frac{2}{3}+\frac{2}{3}$ = ______

5 $\frac{5}{6}+\frac{2}{9}$ = ______

6 $\frac{2}{11}+\frac{3}{9}$ = ______

7 $\frac{7}{8}-\frac{1}{3}$ = ______

8 $\frac{3}{5}-\frac{4}{9}$ = ______

9 $\frac{2}{3}-\frac{2}{9}$ = ______

10 $\frac{7}{8}-\frac{1}{2}$ = ______

11 $\frac{4}{5}-\frac{8}{11}$ = ______

12 $\frac{5}{6}-\frac{5}{11}$ = ______

13 $\frac{2}{3}-\frac{2}{11}$ = ______

14 $\frac{4}{9}-\frac{1}{4}$ = ______

15 $2\frac{1}{3}+4\frac{2}{7}$ = ______

16 $6\frac{1}{8}+5\frac{7}{11}$ = ______

17 $5\frac{3}{10}+3\frac{1}{3}$ = ______

18 $8\frac{1}{11}+\frac{5}{6}$ = ______

19 $7\frac{5}{6}+2\frac{5}{6}$ = ______

20 $5\frac{1}{4}-2\frac{3}{5}$ = ______

21 $6\frac{1}{4}-3\frac{3}{4}$ = ______

22 $7\frac{5}{8}-3\frac{1}{5}$ = ______

23 $7\frac{1}{2}-3\frac{9}{10}$ = ______

24 $8\frac{1}{2}-3\frac{5}{6}$ = ______

25 $7\frac{3}{4}+5\frac{8}{9}-3\frac{3}{4}$ = ______

26 $7\frac{2}{3}-5\frac{1}{2}+3\frac{2}{3}$ = ______

To add or subtract mixed numbers, change them to improper fractions first.

UNIT 32  Stop. Revise. Check.

☞ To add fractions with the same denominator, simply add the numerators.

e.g. (a) $\frac{1}{7} + \frac{3}{7} = \frac{1+3}{7} = \frac{4}{7}$

(b) $\frac{3}{16} + \frac{5}{16} = \frac{3+5}{16} = \frac{8}{16} = \frac{1}{2}$

☞ To subtract fractions with the same denominator, subtract the numerators.

e.g. (a) $\frac{5}{9} - \frac{1}{9} = \frac{5-1}{9} = \frac{4}{9}$

(b) $\frac{7}{8} - \frac{1}{8} = \frac{7-1}{8} = \frac{6}{8} = \frac{3}{4}$

☞ To add or subtract fractions with different denominators, change them to equivalent fractions with the same denominator and then simply add or subtract the numerators.

e.g. (a) $\frac{2}{3} + \frac{4}{5} = \frac{2 \times 5}{3 \times 5} + \frac{4 \times 3}{5 \times 3} = \frac{10}{15} + \frac{12}{15} = \frac{10+12}{15} = \frac{22}{15} = 1\frac{7}{15}$

(b) $\frac{5}{6} - \frac{1}{2} = \frac{5}{6} - \frac{1 \times 3}{2 \times 3} = \frac{5}{6} - \frac{3}{6} = \frac{5-3}{6} = \frac{2}{6} = \frac{1}{3}$

☞ To add or subtract fractions with mixed numbers, change these to improper fractions first and then to equivalent fractions with the same denominator.

e.g. (a) $2\frac{1}{2} + 3\frac{3}{4} = \frac{5}{2} + \frac{15}{4} = \frac{5 \times 2}{2 \times 2} + \frac{15}{4} = \frac{10}{4} + \frac{15}{4} = \frac{25}{4} = 6\frac{1}{4}$

(b) $3\frac{3}{5} - 2\frac{1}{10} = \frac{18}{5} - \frac{21}{10} = \frac{18 \times 2}{5 \times 2} - \frac{21}{10} = \frac{36}{10} - \frac{21}{10} = \frac{15}{10} = \frac{3}{2} = 1\frac{1}{2}$

☞ Reduce the answer to the lowest terms.

☞ A whole number is an improper fraction.

☞ A fraction is also called a rational number.

UNIT 33 Test on Section 3

Instructions: Do not use a calculator.
Attempt all questions.

Time allowed: 25 minutes **Total marks:** 30

	Marks
Question 1: Add the following fractions.	
a $\frac{1}{7} + \frac{2}{7}$ = ______ **b** $\frac{4}{9} + \frac{2}{9}$ = ______	2
c $\frac{2}{9} + \frac{1}{8}$ = ______ **d** $\frac{1}{6} + \frac{3}{5}$ = ______	2
e $\frac{2}{5} + \frac{9}{11}$ = ______ **f** $\frac{5}{11} + \frac{5}{12}$ = ______	2
g $\frac{1}{2} + \frac{5}{9} + \frac{1}{3}$ = ______ **h** $\frac{1}{2} + \frac{3}{4} + \frac{2}{3}$ = ______	2
i $3\frac{3}{4} + 9\frac{1}{2}$ = ______ **j** $3\frac{1}{4} + 8\frac{1}{2}$ = ______	2
k $\frac{7}{12} + 5\frac{1}{2}$ = ______ **l** $3\frac{1}{8} + 5\frac{1}{2}$ = ______	2
Question 2: Subtract the following fractions.	
a $\frac{8}{11} - \frac{3}{11}$ = ______ **b** $\frac{9}{14} - \frac{3}{14}$ = ______	2
c $\frac{4}{9} - \frac{1}{6}$ = ______ **d** $\frac{1}{3} - \frac{2}{9}$ = ______	2
e $\frac{7}{8} - \frac{3}{11}$ = ______ **f** $\frac{12}{13} - \frac{2}{5}$ = ______	2
g $9\frac{3}{4} - 7\frac{1}{2}$ = ______ **h** $6\frac{1}{5} - 4\frac{2}{5}$ = ______	2
i $3\frac{9}{10} - 2\frac{2}{7}$ = ______ **j** $8\frac{6}{7} - 5\frac{1}{2}$ = ______	2
k $8\frac{2}{3} + 2\frac{4}{5} - 1\frac{1}{3}$ = ______ **l** $8\frac{1}{4} - \frac{7}{10} + \frac{2}{5}$ = ______	4
m $\frac{3}{4} + 2\frac{1}{2} - 1\frac{1}{4}$ = ______ **n** $3\frac{3}{7} + 4\frac{1}{2} - 2\frac{3}{4}$ = ______	4
Total marks =	/30

SECTION 4

UNIT 34 Multiplication of Fractions

To multiply fractions, multiply the numerators and multiply the denominators.

Example: Multiply the following fractions.

(a) $\frac{1}{3} \times \frac{1}{3}$ (b) $\frac{2}{5} \times \frac{3}{7}$ (c) $\frac{1}{3} \times \frac{3}{4}$ (d) $\frac{3}{5} \times \frac{2}{5}$

Reduce the answer to its lowest terms.

Solution:

(a) $\frac{1}{3} \times \frac{1}{3} = \frac{1 \times 1}{3 \times 3} = \frac{1}{9}$

(b) $\frac{2}{5} \times \frac{3}{7} = \frac{2 \times 3}{5 \times 7} = \frac{6}{35}$

(c) $\frac{1}{3} \times \frac{3}{4} = \frac{1 \times 3}{3 \times 4} = \frac{3}{12} = \frac{1}{4}$

(d) $\frac{3}{5} \times \frac{2}{5} = \frac{3 \times 2}{5 \times 5} = \frac{6}{25}$

Multiply the following fractions.

1 $\frac{1}{2} \times \frac{1}{2}$ = ______	**2** $\frac{1}{3} \times \frac{1}{4}$ = ______	**3** $\frac{1}{4} \times \frac{1}{5}$ = ______	**4** $\frac{1}{5} \times \frac{1}{6}$ = ______
5 $\frac{2}{3} \times \frac{1}{5}$ = ______	**6** $\frac{3}{4} \times \frac{1}{5}$ = ______	**7** $\frac{2}{3} \times \frac{1}{7}$ = ______	**8** $\frac{3}{5} \times \frac{2}{7}$ = ______
9 $\frac{1}{6} \times \frac{1}{5}$ = ______	**10** $\frac{2}{5} \times \frac{3}{11}$ = ______	**11** $\frac{5}{6} \times \frac{5}{8}$ = ______	**12** $\frac{2}{7} \times \frac{5}{9}$ = ______
13 $\frac{6}{7} \times \frac{3}{5}$ = ______	**14** $\frac{3}{4} \times \frac{5}{13}$ = ______	**15** $\frac{6}{7} \times \frac{4}{5}$ = ______	**16** $\frac{5}{11} \times \frac{7}{8}$ = ______
17 $\frac{3}{17} \times \frac{9}{2}$ = ______	**18** $\frac{3}{5} \times \frac{9}{10}$ = ______	**19** $\frac{8}{9} \times \frac{2}{9}$ = ______	**20** $\frac{6}{11} \times \frac{1}{5}$ = ______
21 $\frac{1}{2} \times \frac{5}{9}$ = ______	**22** $\frac{3}{7} \times \frac{8}{13}$ = ______	**23** $\frac{6}{7} \times \frac{2}{7}$ = ______	**24** $\frac{8}{9} \times \frac{4}{9}$ = ______
25 $\frac{2}{9} \times \frac{5}{3}$ = ______	**26** $\frac{7}{9} \times \frac{5}{8}$ = ______	**27** $\frac{3}{4} \times \frac{9}{8}$ = ______	**28** $\frac{2}{5} \times \frac{7}{9}$ = ______
29 $\frac{23}{25} \times \frac{2}{3}$ = ______	**30** $\frac{7}{13} \times \frac{2}{3}$ = ______	**31** $\frac{8}{9} \times \frac{7}{9}$ = ______	**32** $\frac{6}{7} \times \frac{8}{11}$ = ______
33 $\frac{5}{11} \times \frac{5}{3}$ = ______	**34** $\frac{7}{8} \times \frac{1}{4}$ = ______	**35** $\frac{6}{7} \times \frac{3}{5}$ = ______	**36** $\frac{3}{20} \times \frac{7}{2}$ = ______
37 $\frac{7}{10} \times \frac{3}{10}$ = ______	**38** $\frac{3}{50} \times \frac{1}{2}$ = ______	**39** $\frac{8}{11} \times \frac{6}{5}$ = ______	**40** $\frac{5}{12} \times \frac{7}{8}$ = ______
41 $\frac{6}{17} \times \frac{8}{5}$ = ______	**42** $\frac{5}{6} \times \frac{13}{14}$ = ______	**43** $\frac{5}{7} \times \frac{12}{7}$ = ______	**44** $\frac{8}{9} \times \frac{2}{11}$ = ______
45 $\frac{5}{9} \times \frac{10}{11}$ = ______	**46** $\frac{6}{13} \times \frac{8}{5}$ = ______	**47** $\frac{4}{9} \times \frac{8}{11}$ = ______	**48** $\frac{3}{7} \times \frac{21}{11}$ = ______

UNIT 35 Multiplication of Fractions

We can simplify multiplication of fractions by cancelling (dividing the numerator and the denominator by a common factor) before actually multiplying.

Example: Multiply the following fractions.

(a) $\frac{3}{7} \times \frac{7}{10}$ (b) $\frac{3}{10} \times \frac{1}{9}$ (c) $\frac{21}{50} \times \frac{4}{7}$

Solution:

(a) $\frac{3}{\not{7}_1} \times \frac{\not{7}^1}{10} = \frac{3 \times 1}{1 \times 10} = \frac{3}{10}$

(b) $\frac{{}^1\not{3}}{10} \times \frac{1}{\not{9}_3} = \frac{1 \times 1}{10 \times 3} = \frac{1}{30}$

(c) $\frac{{}^3\not{21}}{\not{50}_{25}} \times \frac{\not{4}^2}{\not{7}_1} = \frac{3 \times 2}{25 \times 1} = \frac{6}{25}$

Multiply the following fractions.

1 $\frac{1}{3} \times \frac{3}{10}$ = ______ ______

2 $\frac{3}{5} \times \frac{5}{7}$ = ______ ______

3 $\frac{7}{10} \times \frac{5}{9}$ = ______ ______

4 $\frac{2}{3} \times \frac{5}{8}$ = ______ ______

5 $\frac{4}{9} \times \frac{5}{6}$ = ______ ______

6 $\frac{10}{13} \times \frac{3}{5}$ = ______ ______

7 $\frac{3}{10} \times \frac{5}{6}$ = ______ ______

8 $\frac{4}{9} \times \frac{18}{26}$ = ______ ______

9 $\frac{5}{9} \times \frac{9}{5}$ = ______ ______

10 $\frac{6}{7} \times \frac{21}{24}$ = ______ ______

11 $\frac{3}{25} \times \frac{10}{12}$ = ______ ______

12 $\frac{8}{15} \times \frac{21}{32}$ = ______ ______

13 $\frac{5}{7} \times \frac{14}{15}$ = ______ ______

14 $\frac{3}{14} \times \frac{21}{9}$ = ______ ______

15 $\frac{7}{8} \times \frac{16}{21}$ = ______ ______

16 $\frac{5}{12} \times \frac{24}{25}$ = ______ ______

17 $\frac{6}{11} \times \frac{22}{36}$ = ______ ______

18 $\frac{2}{9} \times \frac{27}{32}$ = ______ ______

UNIT 36 Multiplication of Fractions

To multiply fractions, multiply the numerators and multiply the denominators and then reduce the answer to the simplest form before multiplying.

Example: Multiply the following fractions.

(a) $\frac{3}{2} \times \frac{2}{5}$ (b) $\frac{5}{6} \times \frac{6}{7}$ (c) $\frac{3}{4} \times \frac{8}{21}$ (d) $\frac{1}{4} \times \frac{2}{3}$

Solution:

(a) $\frac{3}{\cancel{2}_1} \times \frac{\cancel{2}^1}{5} = \frac{3 \times 1}{1 \times 5} = \frac{3}{5}$

(b) $\frac{5}{\cancel{6}_1} \times \frac{\cancel{6}^1}{7} = \frac{5 \times 1}{1 \times 7} = \frac{5}{7}$

(c) $\frac{\cancel{3}^1}{\cancel{4}_1} \times \frac{\cancel{8}^2}{\cancel{21}_3} = \frac{1 \times 2}{1 \times 7} = \frac{2}{7}$

(d) $\frac{1}{\cancel{4}_2} \times \frac{\cancel{2}^1}{3} = \frac{1 \times 1}{2 \times 3} = \frac{1}{6}$

Give the simplest answer for each of the following.

1 $\frac{1}{2} \times \frac{2}{3}$ = ________ **2** $\frac{2}{3} \times \frac{5}{8}$ = ________ **3** $\frac{1}{4} \times \frac{2}{5}$ = ________ **4** $\frac{3}{4} \times \frac{4}{7}$ = ________

5 $\frac{1}{5} \times \frac{5}{7}$ = ________ **6** $\frac{3}{11} \times \frac{11}{13}$ = ________ **7** $\frac{6}{7} \times \frac{7}{13}$ = ________ **8** $\frac{5}{9} \times \frac{9}{7}$ = ________

9 $\frac{3}{7} \times \frac{14}{17}$ = ________ **10** $\frac{8}{9} \times \frac{27}{35}$ = ________ **11** $\frac{5}{9} \times \frac{4}{25}$ = ________ **12** $\frac{3}{11} \times \frac{22}{25}$ = ________

13 $\frac{2}{7} \times \frac{21}{23}$ = ________ **14** $\frac{3}{8} \times \frac{24}{25}$ = ________ **15** $\frac{5}{6} \times \frac{24}{29}$ = ________ **16** $\frac{6}{7} \times \frac{14}{15}$ = ________

17 $\frac{3}{4} \times \frac{8}{9}$ = ________ **18** $\frac{2}{7} \times \frac{7}{4}$ = ________ **19** $\frac{5}{7} \times \frac{14}{15}$ = ________ **20** $\frac{7}{9} \times \frac{18}{7}$ = ________

21 $\frac{2}{9} \times \frac{18}{5}$ = ________ **22** $\frac{2}{13} \times \frac{13}{14}$ = ________ **23** $\frac{20}{27} \times \frac{18}{25}$ = ________ **24** $\frac{6}{35} \times \frac{5}{12}$ = ________

25 $\frac{8}{9} \times \frac{18}{4}$ = ________ **26** $\frac{12}{13} \times \frac{13}{24}$ = ________ **27** $\frac{15}{17} \times \frac{17}{30}$ = ________ **28** $\frac{5}{8} \times \frac{32}{35}$ = ________

UNIT 37 Multiplication of Mixed Numbers

If there are mixed numbers, express them as improper fractions before multiplying.

Example: Multiply the following.

(a) $2\frac{1}{4} \times 1\frac{2}{5}$ (b) $1\frac{1}{4} \times 2\frac{1}{2}$ (c) $2\frac{5}{8} \times 3\frac{1}{2}$

Solution:

(a) $2\frac{1}{4} \times 1\frac{2}{5} = \frac{9}{4} \times \frac{7}{5}$
$= \frac{63}{20} = 3\frac{3}{20}$

(b) $1\frac{1}{4} \times 2\frac{1}{2} = \frac{5}{4} \times \frac{5}{2}$
$= \frac{25}{8} = 3\frac{1}{8}$

(c) $2\frac{5}{8} \times 3\frac{1}{2} = \frac{21}{8} \times \frac{7}{2}$
$= \frac{147}{16} = 9\frac{3}{16}$

Always express your final answer as a mixed number.

Multiply the following.

1 $1\frac{1}{2} \times 2\frac{1}{4}$ = ________ ________

2 $3\frac{1}{5} \times 2\frac{1}{2}$ = ________ ________

3 $4\frac{1}{3} \times 1\frac{1}{2}$ = ________ ________

4 $1\frac{3}{4} \times 1\frac{1}{4}$ = ________ ________

5 $2\frac{1}{4} \times 2\frac{1}{2}$ = ________ ________

6 $3\frac{1}{2} \times 5\frac{1}{4}$ = ________ ________

7 $3\frac{1}{3} \times 2\frac{1}{4}$ = ________ ________

8 $1\frac{2}{5} \times 2\frac{1}{4}$ = ________ ________

9 $3\frac{1}{4} \times 1\frac{1}{2}$ = ________ ________

10 $3\frac{3}{8} \times 4\frac{1}{2}$ = ________ ________

11 $4\frac{4}{5} \times 2\frac{2}{5}$ = ________ ________

12 $3\frac{2}{3} \times 1\frac{2}{5}$ = ________ ________

13 $3\frac{1}{2} \times 2\frac{5}{6}$ = ________ ________

14 $1\frac{1}{2} \times 6\frac{1}{4}$ = ________ ________

15 $2\frac{1}{5} \times 3\frac{3}{10}$ = ________ ________

16 $2\frac{5}{6} \times 3\frac{3}{5}$ = ________ ________

17 $5\frac{1}{2} \times 2\frac{1}{5}$ = ________ ________

18 $4\frac{3}{4} \times 2\frac{7}{8}$ = ________ ________

19 $8\frac{2}{7} \times 2\frac{1}{8}$ = ________ ________

20 $4\frac{2}{3} \times 1\frac{2}{5}$ = ________ ________

21 $12\frac{4}{7} \times 4\frac{1}{2}$ = ________ ________

22 $2\frac{1}{3} \times 2\frac{1}{2}$ = ________ = ________

23 $8\frac{4}{5} \times 2\frac{1}{2}$ = ________ ________

24 $2\frac{1}{2} \times 3\frac{1}{4}$ = ________ ________

25 $5\frac{1}{4} \times 5\frac{1}{2}$ = ________ ________

26 $2\frac{1}{2} \times 2\frac{1}{2}$ = ________ ________

27 $4\frac{3}{4} \times 1\frac{1}{4}$ = ________ ________

28 $8\frac{3}{4} \times 5\frac{2}{5}$ = ________ ________

UNIT 38 Reciprocals

Let us find the product of the following:

(a) $\frac{1}{3} \times \frac{3}{1} = 1$ (b) $\frac{2}{5} \times \frac{5}{2} = 1$ (c) $\frac{9}{10} \times \frac{10}{9} = 1$ (d) $\frac{3}{4} \times \frac{4}{3} = 1$

When two numbers are multiplied and their product is 1, then each number is the **reciprocal** of the other. Remember all numbers (other than zero) have a reciprocal. In order to find a reciprocal, turn the fraction upside down.

To find the reciprocal, turn the fraction upside down.

Example: Find the reciprocal of each of the following numbers.

(a) $\frac{3}{2}$ (b) $2\frac{1}{2}$ (c) 3 (d) $\frac{3}{7}$

Solution:

(a) $\frac{2}{3}$

(b) $2\frac{1}{2} = \frac{5}{2}$
$\therefore$ reciprocal is $\frac{2}{5}$

(c) $3 = \frac{3}{1}$
$\therefore$ reciprocal is $\frac{1}{3}$

(d) $\frac{7}{3} = 2\frac{1}{3}$
$\therefore$ reciprocal is $2\frac{1}{3}$

Find the reciprocal of each of the following numbers.

1	$\frac{1}{4}$	______	**2**	$\frac{9}{4}$	______	**3**	$\frac{10}{7}$	______	**4**	$\frac{15}{4}$	______
5	$\frac{8}{5}$	______	**6**	$\frac{2}{5}$	______	**7**	$\frac{3}{7}$	______	**8**	$\frac{5}{9}$	______
9	$1\frac{1}{2}$	______	**10**	$3\frac{1}{2}$	______	**11**	$4\frac{1}{2}$	______	**12**	$5\frac{1}{4}$	______
13	$\frac{3}{5}$	______	**14**	$\frac{1}{5}$	______	**15**	$4\frac{1}{5}$	______	**16**	$6\frac{1}{4}$	______
17	$\frac{5}{8}$	______	**18**	$\frac{3}{11}$	______	**19**	$\frac{1}{2}$	______	**20**	$\frac{9}{2}$	______
21	$\frac{13}{5}$	______	**22**	9	______	**23**	$\frac{20}{7}$	______	**24**	$9\frac{1}{2}$	______
25	$\frac{5}{6}$	______	**26**	$\frac{2}{7}$	______	**27**	$\frac{9}{10}$	______	**28**	$3\frac{1}{3}$	______
29	$2\frac{1}{4}$	______	**30**	$5\frac{1}{2}$	______	**31**	$\frac{5}{4}$	______	**32**	$\frac{2}{13}$	______
33	$5\frac{3}{4}$	______	**34**	$4\frac{6}{7}$	______	**35**	$3\frac{5}{6}$	______	**36**	$8\frac{3}{4}$	______
37	$1\frac{1}{5}$	______	**38**	$3\frac{5}{7}$	______	**39**	$5\frac{3}{7}$	______	**40**	$9\frac{3}{4}$	______

UNIT 39 Division of Fractions

To divide fractions, multiply the first fraction by the reciprocal of the second fraction.
In other words, to divide fractions invert the second fraction (turn it upside down) and multiply.

Example: Simplify the following fractions.

(a) $\frac{1}{5} \div \frac{1}{7}$ (b) $\frac{2}{3} \div \frac{1}{5}$ (c) $\frac{5}{6} \div \frac{4}{7}$

Solution:

(a) $\frac{1}{5} \div \frac{1}{7} = \frac{1}{5} \times \frac{7}{1} = \frac{7}{5} = 1\frac{2}{5}$

(b) $\frac{2}{3} \div \frac{1}{5} = \frac{2}{3} \times \frac{5}{1} = \frac{10}{3} = 3\frac{1}{3}$

(c) $\frac{5}{6} \div \frac{4}{7} = \frac{5}{6} \times \frac{7}{4} = \frac{35}{24} = 1\frac{11}{24}$

Simplify the following fractions.

1 $\frac{1}{5} \div \frac{1}{3}$ = ____________

2 $\frac{1}{2} \div \frac{1}{7}$ = ____________

3 $\frac{1}{3} \div \frac{1}{2}$ = ____________

4 $\frac{2}{3} \div \frac{5}{4}$ = ____________

5 $\frac{3}{4} \div \frac{4}{5}$ = ____________

6 $\frac{5}{6} \div \frac{6}{5}$ = ____________

7 $\frac{3}{5} \div \frac{5}{2}$ = ____________

8 $\frac{3}{4} \div \frac{2}{5}$ = ____________

9 $\frac{4}{5} \div \frac{1}{3}$ = ____________

10 $\frac{3}{4} \div \frac{8}{11}$ = ____________

11 $\frac{2}{3} \div \frac{3}{4}$ = ____________

12 $\frac{4}{5} \div \frac{7}{8}$ = ____________

13 $\frac{9}{10} \div \frac{10}{11}$ = ____________

14 $\frac{2}{7} \div \frac{1}{11}$ = ____________

15 $\frac{5}{7} \div \frac{7}{9}$ = ____________

16 $\frac{7}{12} \div \frac{4}{5}$ = ____________

17 $\frac{6}{7} \div \frac{7}{8}$ = ____________

18 $\frac{5}{8} \div \frac{3}{11}$ = ____________

19 $\frac{1}{10} \div \frac{1}{11}$ = ____________

20 $\frac{10}{11} \div \frac{11}{8}$ = ____________

21 $\frac{6}{7} \div \frac{5}{3}$ = ____________

22 $\frac{3}{5} \div \frac{1}{2}$ = ____________

23 $\frac{7}{8} \div \frac{1}{5}$ = ____________

24 $\frac{6}{7} \div \frac{5}{9}$ = ____________

UNIT 40 Division of Fractions

The following steps are to be taken when dividing fractions:

Step 1 Change the sign of division to multiplication.
Step 2 Change the second fraction to its reciprocal (turn it upside down).
Step 3 Multiply the numerators together and multiply the denominators together.
Step 4 Simplify the fraction by bringing it down to its lowest form.

When dividing, observe steps 3 and 4 given above.

Example: Simplify the following fractions.

(a) $\frac{3}{4} \div \frac{9}{2}$ (b) $\frac{2}{3} \div \frac{7}{3}$ (c) $\frac{7}{6} \div \frac{14}{4}$

Solution:

(a) $\frac{3}{4} \div \frac{9}{2} = \frac{3}{4} \times \frac{2}{9} = \frac{3 \times 2}{4 \times 9} = \frac{\cancel{6}^{1}}{\cancel{36}_{6}} = \frac{1}{6}$

(b) $\frac{2}{3} \div \frac{7}{3} = \frac{2}{3} \times \frac{3}{7} = \frac{2 \times 3}{3 \times 7} = \frac{\cancel{6}^{2}}{\cancel{21}_{7}} = \frac{2}{7}$

(c) $\frac{7}{6} \div \frac{14}{4} = \frac{7}{6} \times \frac{4}{14} = \frac{7 \times 4}{6 \times 14} = \frac{\cancel{28}^{1}}{\cancel{84}_{3}} = \frac{1}{3}$

Simplify the following fractions.

1 $\frac{3}{10} \div \frac{6}{7} =$ ______ ______

2 $\frac{8}{9} \div \frac{4}{7} =$ ______ ______

3 $\frac{2}{5} \div \frac{4}{15} =$ ______ ______

4 $\frac{6}{7} \div \frac{3}{7} =$ ______ ______

5 $\frac{2}{3} \div \frac{4}{9} =$ ______ ______

6 $\frac{3}{5} \div \frac{6}{10} =$ ______ ______

7 $\frac{2}{9} \div \frac{4}{3} =$ ______ ______

8 $\frac{6}{7} \div \frac{2}{14} =$ ______ ______

9 $\frac{3}{4} \div \frac{6}{7} =$ ______ ______

10 $\frac{3}{7} \div \frac{9}{10} =$ ______ ______

11 $\frac{5}{7} \div \frac{10}{14} =$ ______ ______

12 $\frac{2}{3} \div \frac{8}{9} =$ ______ ______

13 $\frac{3}{14} \div \frac{2}{7} =$ ______ ______

14 $\frac{5}{9} \div \frac{5}{3} =$ ______ ______

15 $\frac{6}{11} \div \frac{3}{2} =$ ______ ______

16 $\frac{5}{8} \div \frac{10}{16} =$ ______ ______

17 $\frac{1}{5} \div \frac{3}{15} =$ ______ ______

18 $\frac{6}{15} \div \frac{2}{3} =$ ______ ______

UNIT 41 Division of Fractions

In division, we turn the question into multiplication and then solve it.
In other words, to divide a fraction multiply it by the reciprocal of the second fraction.

Example: Simplify the following fractions.

(a) $\frac{3}{4} \div \frac{9}{11}$ (b) $\frac{5}{6} \div \frac{25}{27}$ (c) $\frac{8}{9} \div \frac{16}{15}$

Solution:

(a) $\frac{3}{4} \div \frac{9}{11} = \frac{\cancel{3}^{1}}{4} \times \frac{11}{\cancel{9}_{3}} = \frac{11}{12}$

(b) $\frac{5}{6} \div \frac{25}{27} = \frac{\cancel{5}^{1}}{\cancel{6}_{2}} \times \frac{\cancel{27}^{9}}{\cancel{25}_{5}} = \frac{9}{10}$

(c) $\frac{8}{9} \div \frac{16}{15} = \frac{\cancel{8}^{1}}{\cancel{9}_{3}} \times \frac{\cancel{15}^{5}}{\cancel{16}_{2}} = \frac{5}{6}$

Simplify the following fractions.

1 $\frac{8}{9} \div \frac{4}{7}$ = ______

2 $\frac{4}{5} \div \frac{16}{25}$ = ______

3 $\frac{3}{7} \div \frac{9}{14}$ = ______

4 $\frac{7}{10} \div \frac{14}{15}$ = ______

5 $\frac{6}{7} \div \frac{12}{21}$ = ______

6 $\frac{3}{4} \div \frac{6}{18}$ = ______

7 $\frac{3}{5} \div \frac{2}{5}$ = ______

8 $\frac{3}{100} \div \frac{21}{100}$ = ______

9 $\frac{9}{10} \div \frac{3}{20}$ = ______

10 $\frac{5}{9} \div \frac{7}{12}$ = ______

11 $\frac{3}{5} \div \frac{2}{15}$ = ______

12 $\frac{12}{39} \div \frac{4}{13}$ = ______

13 $\frac{4}{9} \div \frac{16}{27}$ = ______

14 $\frac{6}{7} \div \frac{3}{14}$ = ______

15 $\frac{7}{8} \div \frac{15}{16}$ = ______

16 $\frac{2}{11} \div \frac{11}{14}$ = ______

17 $\frac{12}{39} \div \frac{24}{26}$ = ______

18 $\frac{19}{5} \div \frac{19}{10}$ = ______

19 $\frac{8}{15} \div \frac{32}{35}$ = ______

20 $\frac{16}{9} \div \frac{27}{32}$ = ______

21 $\frac{8}{11} \div \frac{22}{24}$ = ______

16 $\frac{6}{13} \div \frac{12}{13}$ = ______

17 $\frac{5}{6} \div \frac{25}{27}$ = ______

18 $\frac{31}{8} \div \frac{31}{16}$ = ______

UNIT 42 Divison of Mixed Numbers

To divide by a fraction, multiply by its reciprocal. When dividing mixed numbers, change the mixed numbers to improper fractions first.

Example: Simplify the following fractions.

(a) $1\frac{3}{8} \div 1\frac{1}{4}$ (b) $3\frac{1}{3} \div 2\frac{1}{2}$ (c) $4\frac{1}{5} \div 2\frac{2}{5}$

To divide mixed numbers, change to improper fractions first.

Solution:

(a) $1\frac{3}{8} \div 1\frac{1}{4} = \frac{11}{8} \div \frac{5}{4}$
$= \frac{11}{{}_2\cancel{8}} \times \frac{{}^1\cancel{4}}{5}$
$= \frac{11}{10} = 1\frac{1}{10}$

(b) $3\frac{1}{3} \div 2\frac{1}{2} = \frac{10}{3} \div \frac{5}{2}$
$= \frac{{}^2\cancel{10}}{3} \times \frac{2}{\cancel{5}_1}$
$= \frac{4}{3} = 1\frac{1}{3}$

(c) $4\frac{1}{5} \div 2\frac{2}{5} = \frac{21}{5} \div \frac{12}{5}$
$= \frac{{}^7\cancel{21}}{{}_1\cancel{5}} \times \frac{{}^1\cancel{5}}{\cancel{12}_4}$
$= \frac{7}{4} = 1\frac{3}{4}$

Simplify the following fractions.

1 $1\frac{1}{2} \div 1\frac{1}{4} =$ ____________

2 $3\frac{3}{4} \div 2\frac{1}{8} =$ ____________

3 $\frac{12}{39} \div 3\frac{1}{4} =$ ____________

4 $4\frac{1}{3} \div 1\frac{4}{15} =$ ____________

5 $3\frac{4}{5} \div 4\frac{4}{5} =$ ____________

6 $12\frac{3}{5} \div 1\frac{1}{2} =$ ____________

7 $8\frac{2}{7} \div 2\frac{1}{12} =$ ____________

8 $2\frac{1}{3} \div 1\frac{2}{3} =$ ____________

9 $1\frac{3}{5} \div 2\frac{2}{5} =$ ____________

10 $2\frac{1}{8} \div 1\frac{3}{4} =$ ____________

11 $1\frac{3}{4} \div 4\frac{1}{3} =$ ____________

12 $3\frac{8}{9} \div 4\frac{5}{9} =$ ____________

13 $2\frac{1}{4} \div \frac{7}{12} =$ ____________

14 $2\frac{1}{2} \div 3\frac{1}{3} =$ ____________

15 $5\frac{1}{2} \div 1\frac{1}{5} =$ ____________

16 $3\frac{2}{3} \div 1\frac{5}{7} =$ ____________

17 $3\frac{1}{2} \div 2\frac{5}{6} =$ ____________

18 $2\frac{1}{4} \div 3\frac{1}{3} =$ ____________

UNIT 43 Further Multiplication and Division of Fractions

So the following steps are to be taken when dividing fractions:

Step 1 Change the sign of division to multiplication.
Step 2 Change the second fraction to its reciprocal (turn it upside down).
Step 3 Cancel the common factors, if any.
Step 4 Solve it by multiplying numerator by numerator and denominator by denominator.

When multiplying or dividing, observe steps 3 and 4 above.

Simplify the following fractions.

1 $\frac{1}{5} \times \frac{1}{5} =$ ________ **2** $\frac{1}{3} \times \frac{1}{7} =$ ________ **3** $\frac{1}{5} \times \frac{1}{6} =$ ________

4 $\frac{2}{3} \times \frac{3}{7} =$ ________ **5** $\frac{5}{6} \times \frac{12}{13} =$ ________ **6** $\frac{8}{9} \times \frac{12}{13} =$ ________

7 $\frac{5}{7} \times \frac{5}{7} =$ ________ **8** $\frac{7}{9} \times \frac{5}{6} =$ ________ **9** $\frac{8}{9} \times \frac{12}{14} =$ ________

10 $\frac{2}{3} \times \frac{5}{8} =$ ________ **11** $\frac{4}{5} \times \frac{1}{4} =$ ________ **12** $\frac{2}{3} \times \frac{3}{8} =$ ________

13 $\frac{9}{10} \times \frac{20}{21} =$ ________ **14** $\frac{12}{15} \times \frac{5}{36} =$ ________ **15** $\frac{24}{35} \times \frac{14}{36} =$ ________

16 $2\frac{3}{5} \times 1\frac{1}{2} =$ ________ **17** $3\frac{4}{7} \times \frac{5}{8} =$ ________ **18** $8\frac{1}{4} \times 1\frac{1}{4} =$ ________

19 $\frac{1}{3} \div \frac{1}{4} =$ ________ **20** $\frac{1}{5} \div \frac{1}{6} =$ ________ **21** $\frac{1}{8} \div \frac{1}{11} =$ ________

22 $\frac{1}{2} \div \frac{5}{2} =$ ________ **23** $\frac{4}{9} \div \frac{8}{27} =$ ________ **24** $\frac{5}{6} \div \frac{10}{12} =$ ________

25 $\frac{2}{3} \div \frac{1}{3} =$ ________ **26** $\frac{5}{6} \div \frac{2}{5} =$ ________ **27** $\frac{11}{13} \div \frac{22}{13} =$ ________

UNIT 44 

Quick Questions

This unit of 'Quick Questions' is designed to make sure that you think fast and remain in touch with the previous work of the section.

An improper fraction is when the numerator is equal to or greater than the denominator.

Simplify the following.

1 $\frac{1}{2} \times \frac{1}{2}$ = ______ **2** $\frac{1}{3} \times \frac{1}{3}$ = ______ **3** $\frac{1}{4} \times \frac{1}{4}$ = ______

4 $\frac{2}{3} \times \frac{1}{5}$ = ______ **5** $\frac{3}{4} \times \frac{1}{5}$ = ______ **6** $\frac{5}{6} \times \frac{1}{3}$ = ______

7 $\frac{2}{3} \times \frac{2}{5}$ = ______ **8** $\frac{3}{5} \times \frac{2}{7}$ = ______ **9** $\frac{4}{7} \times \frac{3}{5}$ = ______

10 $\frac{7}{9} \times \frac{3}{4}$ = ______ **11** $\frac{8}{9} \times \frac{3}{4}$ = ______ **12** $\frac{6}{7} \times \frac{14}{18}$ = ______

13 $\frac{4}{5} \times \frac{5}{9}$ = ______ **14** $\frac{9}{10} \times \frac{5}{12}$ = ______ **15** $\frac{9}{11} \times \frac{22}{27}$ = ______

16 $\frac{5}{8} \times 64$ = ______ **17** $\frac{3}{5} \times 25$ = ______ **18** $\frac{5}{7} \times 56$ = ______

19 $1\frac{3}{4} \times 2\frac{1}{2}$ = ______ **20** $5\frac{1}{3} \times 1\frac{1}{2}$ = ______ **21** $6\frac{5}{6} \times \frac{3}{5}$ = ______

22 $\frac{1}{2} \div \frac{1}{2}$ = ______ **23** $\frac{1}{3} \div \frac{1}{3}$ = ______ **24** $\frac{1}{4} \div \frac{1}{4}$ = ______

25 $\frac{3}{4} \div \frac{1}{2}$ = ______ **26** $\frac{5}{6} \div \frac{1}{2}$ = ______ **27** $\frac{3}{8} \div \frac{1}{4}$ = ______

28 $\frac{3}{5} \div \frac{2}{15}$ = ______ **29** $\frac{1}{9} \div \frac{1}{12}$ = ______ **30** $\frac{2}{9} \div \frac{1}{3}$ = ______

31 $\frac{2}{5} \div \frac{1}{10}$ = ______ **32** $\frac{3}{2} \div \frac{6}{8}$ = ______ **33** $\frac{5}{7} \div \frac{10}{14}$ = ______

34 $\frac{1}{6} \div \frac{2}{3}$ = ______ **35** $\frac{5}{8} \div \frac{4}{5}$ = ______ **36** $\frac{7}{9} \div \frac{9}{14}$ = ______

37 $\frac{3}{8} \div \frac{6}{10}$ = ______ **38** $\frac{6}{11} \div \frac{1}{11}$ = ______ **39** $\frac{5}{13} \div \frac{8}{13}$ = ______

40 $\frac{6}{25} \div \frac{12}{25}$ = ______ **41** $\frac{3}{14} \div \frac{3}{7}$ = ______ **42** $8 \div \frac{4}{7}$ = ______

43 $1\frac{1}{2} \div 1\frac{1}{2}$ = ______ **44** $2\frac{1}{2} \div 2\frac{1}{2}$ = ______ **45** $2\frac{1}{5} \div 3\frac{1}{5}$ = ______

UNIT 45

Stay in Touch

The unit 'Stay In Touch' is intended to keep you in touch with what you have done before. It also ensures that all types of questions in this section receive constant revision.

Change the following improper fractions to mixed numbers.

1 $\frac{9}{2}$ = ______ **2** $\frac{13}{3}$ = ______ **3** $\frac{11}{2}$ = ______ **4** $\frac{25}{4}$ = ______

5 $\frac{35}{6}$ = ______ **6** $\frac{39}{5}$ = ______ **7** $\frac{72}{10}$ = ______ **8** $\frac{48}{7}$ = ______

Change the following mixed numbers to improper fractions.

9 $3\frac{1}{4}$ = ______ **10** $5\frac{5}{6}$ = ______ **11** $9\frac{2}{3}$ = ______ **12** $11\frac{5}{6}$ = ______

13 $8\frac{2}{3}$ = ______ **14** $9\frac{5}{7}$ = ______ **15** $11\frac{1}{3}$ = ______ **16** $7\frac{6}{7}$ = ______

Simplify the following.

17 $\frac{10}{45}$ = ______ **18** $\frac{8}{64}$ = ______ **19** $\frac{12}{50}$ = ______ **20** $\frac{30}{58}$ = ______

Complete the following equivalent fractions.

21 $\frac{3}{10} = \frac{12}{\square}$ **22** $\frac{5}{7} = \frac{\square}{35}$ **23** $\frac{9}{11} = \frac{27}{\square}$ **24** $\frac{15}{20} = \frac{3}{\square}$

25 $\frac{16}{20} = \frac{\square}{5}$ **26** $\frac{36}{81} = \frac{\square}{9}$ **27** $\frac{45}{60} = \frac{3}{\square}$ **28** $\frac{30}{70} = \frac{\square}{7}$

Arrange each set of fractions in ascending order.

29 $\frac{1}{5}, \frac{4}{5}, \frac{2}{5}$ ______ **30** $\frac{1}{2}, \frac{2}{3}, \frac{1}{4}$ ______ **31** $\frac{1}{3}, \frac{1}{4}, \frac{2}{3}$ ______

32 $\frac{3}{4}, \frac{5}{6}, \frac{1}{2}$ ______ **33** $\frac{8}{12}, \frac{1}{4}, \frac{7}{8}$ ______ **34** $\frac{2}{3}, \frac{5}{6}, \frac{1}{3}$ ______

Simplify the following.

35 $\frac{1}{9} + \frac{4}{9}$ = ______ **36** $\frac{3}{8} + \frac{3}{8}$ = ______ **37** $\frac{5}{6} - \frac{2}{3}$ = ______

38 $8\frac{1}{4} - 2\frac{1}{2}$ = ______ **39** $5\frac{1}{6} + \frac{2}{5}$ = ______ **40** $4\frac{1}{2} - 1\frac{3}{4}$ = ______

UNIT 46 Stop. Revise. Check.

☞ To multiply fractions, multiply the numerators and multiply the denominators.

e.g. (a) $\frac{1}{3} \times \frac{5}{7} = \frac{1 \times 5}{3 \times 7} = \frac{5}{21}$ (b) $\frac{2}{9} \times \frac{4}{5} = \frac{2 \times 4}{9 \times 5} = \frac{8}{45}$

☞ We can simplify multiplication of fractions by cancelling (dividing the numerator and denominator by the common factor) before actually multiplying.

e.g. (a) $\frac{3}{\cancel{7}_1} \times \frac{\cancel{7}^1}{8} = \frac{3 \times 1}{1 \times 8} = \frac{3}{8}$ (b) $\frac{12}{\cancel{18}_1} \times \frac{\cancel{36}^2}{13} = \frac{12 \times 2}{13} = \frac{24}{13} = 1\frac{11}{13}$

☞ If there are mixed numbers, express these as improper fractions before multiplying.

e.g. (a) $2\frac{1}{2} \times 1\frac{1}{4} = \frac{5}{2} \times \frac{5}{4} = \frac{25}{8} = 3\frac{1}{8}$ (b) $1\frac{1}{3} \times 2\frac{1}{2} = \frac{4}{3} \times \frac{5}{2} = \frac{20}{6} = \frac{10}{3} = 3\frac{1}{3}$

☞ When the product of two numbers is 1, then each is the reciprocal of the other.

e.g. $\frac{2}{3} \times \frac{3}{2} = 1$ $\therefore$ $\frac{2}{3}$ is the reciprocal of $\frac{3}{2}$

☞ To find a reciprocal, turn the fraction upside down. e.g. the reciprocal of $\frac{5}{7}$ is $\frac{7}{5}$

☞ To divide a fraction, multiply it by the reciprocal of the second fraction.

e.g. $\frac{8}{9} \div \frac{2}{7} = \frac{\cancel{8}^4}{9} \times \frac{7}{\cancel{2}_1} = \frac{4 \times 7}{9} = \frac{28}{9} = 3\frac{1}{9}$

☞ When dividing mixed numbers, change the mixed numbers to improper fractions first.

e.g. $2\frac{3}{4} \div 3\frac{1}{3} = \frac{11}{4} \div \frac{10}{3} = \frac{11}{4} \times \frac{3}{10} = \frac{33}{40}$

UNIT 47 Test on Section 4

Instructions: Do not use a calculator.
Attempt all questions.

Time allowed: 25 minutes **Total marks:** 26

		Marks
Question 1: Multiply the following fractions.		
a $\frac{1}{6} \times \frac{1}{6}$ = ______	**b** $\frac{2}{5} \times \frac{6}{7}$ = ______	2
c $\frac{8}{9} \times \frac{3}{4}$ = ______	**d** $\frac{15}{28} \times \frac{1}{2}$ = ______	2
e $\frac{4}{7} \times \frac{3}{5}$ = ______	**f** $\frac{8}{9} \times \frac{12}{20}$ = ______	2
g $\frac{9}{20} \times \frac{5}{18}$ = ______	**h** $\frac{6}{7} \times \frac{14}{24}$ = ______	2
i $2\frac{1}{3} \times 1\frac{1}{3}$ = ______	**j** $\frac{9}{10} \times \frac{3}{20}$ = ______	2
k $\frac{7}{9} \times \frac{18}{28}$ = ______	**l** $3\frac{2}{7} \times 1\frac{1}{6}$ = ______	2
Question 2: Simplify the following fractions.		
a $\frac{5}{8} \div \frac{1}{2}$ = ______	**b** $\frac{6}{7} \div \frac{12}{14}$ = ______	2
c $\frac{3}{7} \div \frac{9}{14}$ = ______	**d** $\frac{1}{6} \div \frac{2}{3}$ = ______	2
e $\frac{8}{25} \div \frac{16}{35}$ = ______	**f** $\frac{5}{14} \div \frac{7}{5}$ = ______	2
g $3\frac{1}{2} \div 3\frac{1}{2}$ = ______	**h** $10 \div \frac{70}{80}$ = ______	2
i $\frac{8}{15} \div \frac{9}{35}$ = ______	**j** $5\frac{5}{8} \div 4\frac{1}{4}$ = ______	2
k $\frac{7}{18} \div \frac{21}{36}$ = ______	**l** $\frac{11}{25} \div \frac{121}{100}$ = ______	2
m $3\frac{2}{5} \div 1\frac{1}{5}$ = ______	**n** $9\frac{7}{8} \div \frac{1}{8}$ = ______	2
	Total marks =	/26

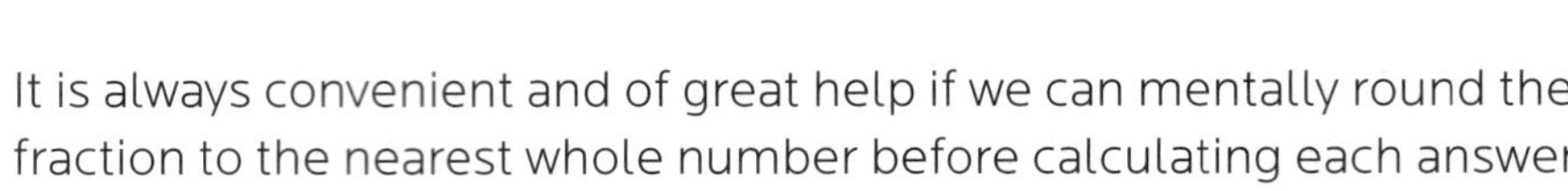

SECTION 5

UNIT 48 Rounding Fractions

It is always convenient and of great help if we can mentally round the fraction to the nearest whole number before calculating each answer.

Example: Round the following to the nearest whole number.

(a) $2\frac{3}{4}$ (b) $5\frac{1}{6}$ (c) $4\frac{1}{3}$

Solution:

(a) In fraction $2\frac{3}{4}$, since $\frac{3}{4} > \frac{1}{2}$, round it up. $\therefore$ $2\frac{3}{4}$ to the nearest whole number is 3.

(b) Since $\frac{1}{6} < \frac{1}{2}$, round it down. $5\frac{1}{6}$ to the nearest whole number = 5.

(c) Since $\frac{1}{3} < \frac{1}{2}$, round it down. $4\frac{1}{3}$ = 4 to the nearest whole number.

Round to the nearest whole number. (Remember: $\geq \frac{1}{2}$ round up; $< \frac{1}{2}$ round down.)

1 $\frac{5}{6}$ ______	**2** $2\frac{1}{2}$ ______	**3** $3\frac{2}{3}$ ______	**4** $5\frac{1}{8}$ ______
5 $6\frac{6}{7}$ ______	**6** $8\frac{1}{4}$ ______	**7** $7\frac{1}{5}$ ______	**8** $3\frac{5}{8}$ ______
9 $2\frac{4}{11}$ ______	**10** $5\frac{5}{8}$ ______	**11** $6\frac{3}{7}$ ______	**12** $4\frac{3}{12}$ ______
13 $\frac{81}{100}$ ______	**14** $61\frac{5}{87}$ ______	**15** $7\frac{3}{100}$ ______	**16** $5\frac{68}{100}$ ______
17 $5\frac{681}{1000}$ ______	**18** $2\frac{123}{1000}$ ______	**19** $6\frac{39}{1000}$ ______	**20** $9\frac{63}{100}$ ______
21 $8\frac{81}{100}$ ______	**22** $62\frac{7}{100}$ ______	**23** $15\frac{53}{100}$ ______	**24** $12\frac{73}{100}$ ______

Round the following to the nearest whole number mentally and give an approximate answer before calculating each answer exactly.

25 $3\frac{1}{4} + 2\frac{1}{2}$ = ______	**26** $5\frac{3}{4} + 1\frac{1}{4}$ = ______	**27** $5\frac{1}{2} + 2\frac{3}{4}$ = ______
28 $8\frac{5}{6} + 1\frac{1}{2}$ = ______	**29** $3\frac{3}{4} + 2\frac{1}{8}$ = ______	**30** $9\frac{1}{2} + 2\frac{7}{10}$ = ______
31 $2\frac{8}{9} + 3\frac{1}{3}$ = ______	**32** $7\frac{1}{9} - 2\frac{3}{4}$ = ______	**33** $15\frac{6}{7} - 10\frac{2}{3}$ = ______
34 $8\frac{1}{6} - 5\frac{1}{4}$ = ______	**35** $15\frac{2}{3} - 12\frac{1}{3}$ = ______	**36** $8\frac{53}{100} - 5\frac{7}{10}$ = ______
37 $4\frac{1}{2} + 2\frac{1}{4} - 1\frac{1}{2}$ = ______	**38** $6\frac{1}{4} + 3\frac{3}{4} - 2\frac{1}{4}$ = ______	**39** $8\frac{5}{6} + \frac{1}{12} - 5\frac{3}{4}$ = ______

UNIT 49 Fractions of Whole Numbers

Fractions are part of a whole number.

Sometimes the word 'of' is used for multiplication.

Example: Find the following.

(a) $\frac{1}{4}$ of 12 (b) $\frac{1}{2}$ of 20 (c) $\frac{3}{4}$ of 60 (d) $\frac{5}{6}$ of 30

Solution:

(a) $\frac{1}{4}$ of 12 $= \frac{1}{4} \times 12 = 3$

(b) $\frac{1}{2}$ of 20 $= \frac{1}{2} \times 20 = 10$

(c) $\frac{3}{4}$ of 60 $= \frac{3}{\cancel{4}_1} \times \cancel{60}^{15} = 45$

(d) $\frac{5}{6}$ of 30 $= \frac{5}{\cancel{6}_1} \times \cancel{30}^{5} = 25$

Find the following.

1 $\frac{1}{3}$ of 6 ______	**2** $\frac{1}{4}$ of 8 ______	**3** $\frac{1}{2}$ of 12 ______	**4** $\frac{3}{4}$ of 16 ______
5 $\frac{1}{5}$ of 5 ______	**6** $\frac{1}{6}$ of 12 ______	**7** $\frac{1}{9}$ of 9 ______	**8** $\frac{1}{10}$ of 20 ______
9 $\frac{2}{3}$ of 30 ______	**10** $\frac{1}{2}$ of 24 ______	**11** $\frac{1}{10}$ of 60 ______	**12** $\frac{2}{7}$ of 14 ______
13 $\frac{5}{9}$ of 90 ______	**14** $\frac{2}{3}$ of 24 ______	**15** $\frac{1}{2}$ of 80 ______	**16** $\frac{1}{3}$ of 18 ______
17 $\frac{3}{4}$ of 120 ______	**18** $\frac{1}{4}$ of 88 ______	**19** $\frac{1}{2}$ of 72 ______	**20** $\frac{2}{3}$ of 27 ______
21 $\frac{2}{5}$ of 500 ______	**22** $\frac{3}{4}$ of 24 ______	**23** $\frac{5}{11}$ of 33 ______	**24** $\frac{4}{5}$ of 500 ______
25 $\frac{3}{8}$ of 64 ______	**26** $\frac{1}{7}$ of 77 ______	**27** $\frac{1}{5}$ of 25 ______	**28** $\frac{4}{9}$ of 180 ______
29 $\frac{1}{2}$ of 40 ______	**30** $\frac{4}{9}$ of 36 ______	**31** $\frac{5}{11}$ of 880 ______	**32** $\frac{1}{2}$ of 800 ______
33 $\frac{1}{3}$ of 300 ______	**34** $\frac{1}{2}$ of 700 ______	**35** $\frac{1}{6}$ of 1200 ______	**36** $\frac{2}{5}$ of 200 ______
37 $\frac{2}{3}$ of 300 ______	**38** $\frac{5}{6}$ of 600 ______	**39** $\frac{5}{7}$ of 770 ______	**40** $\frac{1}{3}$ of 270 ______
41 $\frac{2}{5}$ of 75 ______	**42** $\frac{1}{3}$ of 240 ______	**43** $\frac{1}{10}$ of 200 ______	**44** $\frac{1}{8}$ of 72 ______
45 $\frac{2}{5}$ of 15 ______	**46** $\frac{5}{8}$ of 800 ______	**47** $\frac{5}{6}$ of 66 ______	**48** $\frac{2}{9}$ of 270 ______

UNIT 50 Squares of Fractions

To find the reciprocal turn the fraction upside down.

Example: Find the following.

(a) $\frac{1}{2} \times \frac{1}{2}$ (b) $\frac{2}{3} \times \frac{2}{3}$ (c) $\left(\frac{3}{4}\right)^2$ (d) $\left(\frac{5}{2}\right)^2$

Solution:

(a) $\frac{1}{2} \times \frac{1}{2} = \frac{1 \times 1}{2 \times 2} = \frac{1}{4}$

(b) $\frac{2}{3} \times \frac{2}{3} = \frac{2 \times 2}{3 \times 3} = \frac{4}{9}$

(c) $\left(\frac{3}{4}\right)^2 = \frac{3}{4} \times \frac{3}{4} = \frac{3 \times 3}{4 \times 4} = \frac{9}{16}$

(d) $\left(\frac{5}{2}\right)^2 = \frac{5}{2} \times \frac{5}{2} = \frac{5 \times 5}{2 \times 2} = \frac{25}{4} = 6\frac{1}{4}$

Find the following.

1 $\frac{1}{5} \times \frac{1}{5}$ = ______ **2** $\frac{1}{3} \times \frac{1}{3}$ = ______ **3** $\frac{1}{4} \times \frac{1}{4}$ = ______ **4** $\frac{1}{6} \times \frac{1}{6}$ = ______

5 $\frac{2}{5} \times \frac{2}{5}$ = ______ **6** $\frac{3}{5} \times \frac{3}{5}$ = ______ **7** $\frac{4}{7} \times \frac{4}{7}$ = ______ **8** $\frac{3}{8} \times \frac{3}{8}$ = ______

9 $\frac{4}{9} \times \frac{4}{9}$ = ______ **10** $\frac{5}{7} \times \frac{5}{7}$ = ______ **11** $\frac{6}{11} \times \frac{6}{11}$ = ______ **12** $\frac{7}{9} \times \frac{7}{9}$ = ______

13 $\frac{2}{7} \times \frac{2}{7}$ = ______ **14** $\frac{3}{8} \times \frac{3}{8}$ = ______ **15** $\frac{5}{9} \times \frac{5}{9}$ = ______ **16** $\frac{6}{7} \times \frac{6}{7}$ = ______

17 $1\frac{1}{2} \times 1\frac{1}{2}$ = ______ **18** $2\frac{1}{2} \times 2\frac{1}{2}$ = ______ **19** $2\frac{1}{4} \times 2\frac{1}{4}$ = ______ **20** $3\frac{1}{2} \times 3\frac{1}{2}$ = ______

Find the following squares.

21 $\left(\frac{1}{2}\right)^2$ = ______ **22** $\left(\frac{1}{3}\right)^2$ = ______ **23** $\left(\frac{1}{4}\right)^2$ = ______ **24** $\left(\frac{1}{5}\right)^2$ = ______

25 $\left(\frac{2}{3}\right)^2$ = ______ **26** $\left(\frac{2}{5}\right)^2$ = ______ **27** $\left(\frac{2}{7}\right)^2$ = ______ **28** $\left(\frac{2}{9}\right)^2$ = ______

29 $\left(\frac{3}{5}\right)^2$ = ______ **30** $\left(\frac{3}{7}\right)^2$ = ______ **31** $\left(\frac{4}{7}\right)^2$ = ______ **32** $\left(\frac{5}{7}\right)^2$ = ______

33 $\left(\frac{1}{8}\right)^2$ = ______ **34** $\left(\frac{3}{8}\right)^2$ = ______ **35** $\left(\frac{5}{8}\right)^2$ = ______ **36** $\left(\frac{7}{8}\right)^2$ = ______

37 $\left(\frac{1}{9}\right)^2$ = ______ **38** $\left(\frac{2}{9}\right)^2$ = ______ **39** $\left(\frac{3}{10}\right)^2$ = ______ **40** $\left(\frac{9}{10}\right)^2$ = ______

41 $\left(1\frac{1}{2}\right)^2$ = ______ **42** $\left(2\frac{1}{2}\right)^2$ = ______ **43** $\left(3\frac{1}{2}\right)^2$ = ______ **44** $\left(5\frac{1}{3}\right)^2$ = ______

UNIT 51 Square Roots of Fractions

Example: Find the square root of each of the following.

(a) $\frac{1}{9}$ (b) $\frac{1}{16}$ (c) $\frac{9}{49}$ (d) $\frac{4}{25}$

Solution:

(a) $\sqrt{\frac{1}{9}} = \frac{\sqrt{1}}{\sqrt{9}} = \frac{1}{3}$ (b) $\sqrt{\frac{1}{16}} = \frac{\sqrt{1}}{\sqrt{16}} = \frac{1}{4}$ (c) $\sqrt{\frac{9}{49}} = \frac{\sqrt{9}}{\sqrt{49}} = \frac{3}{7}$ (d) $\sqrt{\frac{4}{25}} = \frac{\sqrt{4}}{\sqrt{25}} = \frac{2}{5}$

Find the square root of each of the following.

1 $\frac{1}{4}$ = ______	**2** $\frac{1}{9}$ = ______	**3** $\frac{1}{16}$ = ______	**4** $\frac{1}{25}$ = ______
5 $\frac{1}{36}$ = ______	**6** $\frac{1}{49}$ = ______	**7** $\frac{1}{64}$ = ______	**8** $\frac{1}{81}$ = ______
9 $\frac{4}{9}$ = ______	**10** $\frac{9}{16}$ = ______	**11** $\frac{16}{25}$ = ______	**12** $\frac{16}{36}$ = ______
13 $\frac{9}{64}$ = ______	**14** $\frac{25}{36}$ = ______	**15** $\frac{36}{49}$ = ______	**16** $\frac{25}{81}$ = ______
17 $\frac{16}{100}$ = ______	**18** $\frac{25}{121}$ = ______	**19** $\frac{36}{144}$ = ______	**20** $\frac{4}{81}$ = ______
21 $\frac{9}{121}$ = ______	**22** $\frac{100}{144}$ = ______	**23** $\frac{121}{625}$ = ______	**24** $\frac{144}{225}$ = ______

Find the values of the following.

25 $\sqrt{\frac{4}{9}}$ = ______	**26** $\sqrt{\frac{16}{81}}$ = ______	**27** $\sqrt{\frac{9}{25}}$ = ______	**28** $\sqrt{\frac{36}{49}}$ = ______
29 $\sqrt{\frac{64}{81}}$ = ______	**30** $\sqrt{\frac{16}{49}}$ = ______	**31** $\sqrt{\frac{25}{100}}$ = ______	**32** $\sqrt{\frac{121}{225}}$ = ______
33 $\sqrt{\frac{4}{36}}$ = ______	**34** $\sqrt{\frac{9}{36}}$ = ______	**35** $\sqrt{\frac{16}{64}}$ = ______	**36** $\sqrt{\frac{25}{625}}$ = ______
37 $\sqrt{\frac{4}{25}}$ = ______	**38** $\sqrt{\frac{9}{64}}$ = ______	**39** $\sqrt{\frac{9}{81}}$ = ______	**40** $\sqrt{\frac{16}{100}}$ = ______
41 $\sqrt{\frac{25}{81}}$ = ______	**42** $\sqrt{\frac{25}{121}}$ = ______	**43** $\sqrt{\frac{25}{144}}$ = ______	**44** $\sqrt{\frac{36}{121}}$ = ______
45 $\sqrt{\frac{36}{169}}$ = ______	**46** $\sqrt{\frac{49}{121}}$ = ______	**47** $\sqrt{\frac{64}{144}}$ = ______	**48** $\sqrt{\frac{81}{169}}$ = ______

UNIT 52 Order of Operations

In life, the order in which we do things is very important. For example, we put shoes on socks, not socks on shoes. The same is true in mathematics. The following rules should be observed when dealing with a problem consisting of more than one operation.

1. Grouping symbols (brackets) should always be calculated first.
2. Multiplication or division should be calculated next, whichever comes first when working from left to right.
3. Addition or subtraction should be calculated next, working from left to right.

To add or subtract mixed numbers, write them as improper fractions first.

Example: Simplify the following.

(a) $\left(2\frac{3}{4}+\frac{1}{4}\right)-2\frac{1}{2}$ (b) $5\frac{5}{6}-\left(3\frac{2}{3}-\frac{1}{6}\right)$ (c) $\frac{8}{9}+\frac{2}{3}\times\frac{1}{4}$

Solution:

(a) $\left(2\frac{3}{4}+\frac{1}{4}\right)-2\frac{1}{2}$
$=3-2\frac{1}{2}$
$=\frac{1}{2}$

(b) $5\frac{5}{6}-\left(3\frac{2}{3}-\frac{1}{6}\right)$
$=5\frac{5}{6}-3\frac{1}{2}$
$=2\frac{1}{3}$

(c) $\frac{8}{9}+\frac{2}{3}\times\frac{1}{4}$
$=\frac{8}{9}+\frac{1}{6}$
$=\frac{19}{18}=1\frac{1}{18}$

Use the order of operations to simplify the following.

1 $\frac{3}{4}+\frac{1}{2}\times\frac{1}{3}$ = ______

2 $\frac{2}{3}\times\frac{9}{10}+\frac{1}{5}\times\frac{5}{6}$ = ______

3 $3\frac{5}{8}\div\left(\frac{1}{2}+\frac{1}{4}\right)$ = ______

4 $\frac{4}{7}+\frac{5}{9}\div\frac{10}{3}$ = ______

5 $4\frac{1}{2}+\left(1\frac{1}{2}\div\frac{1}{2}\right)$ = ______

6 $2\frac{8}{15}-\frac{4}{9}\div\frac{2}{3}+\frac{1}{5}$ = ______

7 $2\frac{1}{2}\div1\frac{1}{2}+\frac{3}{4}$ = ______

8 $3\frac{3}{4}+2\frac{5}{6}\times\frac{12}{17}$ = ______

9 $\left(5\frac{1}{4}+2\frac{1}{2}\right)\times3\frac{1}{2}$ = ______

10 $\frac{5}{8}+\frac{3}{4}\times\frac{1}{6}+\frac{7}{8}$ = ______

11 $7\frac{5}{7}-\frac{3}{8}\times\frac{1}{7}$ = ______

12 $\frac{1}{2}\times\frac{1}{5}\div\frac{1}{25}+\frac{2}{5}$ = ______

13 $\frac{5}{16}+\frac{4}{5}\div\frac{2}{3}+\frac{7}{8}$ = ______

14 $\frac{9}{11}+\frac{1}{2}\div\frac{1}{5}+\frac{7}{22}$ = ______

UNIT 53 Order of Operations

There is a very helpful rule for the order of operations—we should work in the order given by the word **BODMAS**.

B O D M A S

Brackets, of, Division, Multiplication, Addition, Subtraction

Example: Simplify the following.

(a) $\frac{5}{6}+\frac{1}{2}\times\frac{1}{3}$ (b) $\left(\frac{1}{3}+\frac{1}{4}\right)+\frac{2}{3}\div\frac{8}{5}$ (c) $1\frac{7}{8}-\left(\frac{3}{4}+\frac{3}{2}\times\frac{1}{4}\right)$

Solution:

(a) $\frac{5}{6}+\frac{1}{2}\times\frac{1}{3}$

$=\frac{5}{6}+\frac{1}{6}$

$=\frac{6}{6}$

$=1$

(b) $\left(\frac{1}{3}+\frac{1}{4}\right)+\frac{2}{3}\div\frac{8}{5}$

$=\left(\frac{4}{12}+\frac{3}{12}\right)+\frac{2}{3}\div\frac{8}{5}$

$=\frac{7}{12}+\frac{\cancel{2}^{1}}{3}\times\frac{5}{\cancel{8}_{4}}$

$=\frac{7}{12}+\frac{5}{12}$

$=\frac{12}{12}$

$=1$

(c) $1\frac{7}{8}-\left(\frac{3}{4}+\frac{3}{2}\times\frac{1}{4}\right)$

$=\frac{15}{8}-\left(\frac{3}{4}+\frac{3}{8}\right)$

$=\frac{15}{8}-\frac{9}{8}$

$=\frac{15-9}{8}$

$=\frac{6}{8}$

$=\frac{3}{4}$

Use the BODMAS rule to simplify the following.

1 $\frac{5}{6}-\frac{1}{2}\times\frac{1}{3}$ ______

2 $\frac{3}{4}\times\frac{4}{5}+\frac{3}{5}\times\frac{1}{3}$ ______

3 $5\frac{1}{4}\div\left(\frac{1}{2}-\frac{1}{4}\right)$ ______

4 $\frac{3}{8}+\frac{4}{9}\div\frac{8}{3}$ ______

5 $6\frac{1}{4}+\left(2\frac{3}{4}-\frac{3}{4}\right)$ ______

6 $\frac{6}{7}-\frac{3}{5}\div\frac{9}{10}+3\frac{1}{3}$ ______

7 $5\frac{6}{7}-\frac{3}{8}\div\frac{7}{8}$ ______

8 $\frac{3}{7}\times\frac{14}{9}+\frac{1}{2}\div\frac{4}{5}$ ______

9 $3\frac{1}{2}\times\left(\frac{1}{2}+\frac{2}{3}\right)\div\frac{3}{4}$ ______

UNIT 54 Fractions of Quantities

Fractions are commonly and constantly used in our day-to-day life to show the parts of a whole amount or a quantity.

Example: Find the following amounts.

(a) $\frac{1}{4}$ of \$80 (b) $\frac{2}{3}$ of 12 hours (c) $\frac{3}{5}$ of 25 km (d) $\frac{9}{11}$ of \$66

Solution:

(a) $\frac{1}{\cancel{4}_1} \times \cancel{\$80}_{20}$ $= \$20$

(b) $\frac{2}{\cancel{3}_1} \times \cancel{12}_4$ $= 8$ hours

(c) $\frac{3}{\cancel{5}_1} \times \cancel{25}_5$ $= 15$ km

(d) $\frac{9}{\cancel{11}_1} \times \cancel{\$66}_6$ $= \$54$

Find the value of the following.

1 $\frac{1}{2}$ of \$8 = ______	**2** $\frac{2}{5}$ of \$25 = ______	**3** $\frac{1}{4}$ of \$32 = ______
4 $\frac{1}{3}$ of \$66 = ______	**5** $\frac{3}{4}$ of \$100 = ______	**6** $\frac{5}{6}$ of 1 hour = ______
7 $\frac{1}{5}$ of 20 min = ______	**8** $\frac{9}{10}$ of 1 kg = ______	**9** $\frac{9}{10}$ of a century = ______
10 $\frac{3}{10}$ of 1 metre = ______	**11** $\frac{1}{3}$ of 21 weeks = ______	**12** $\frac{3}{5}$ of 1 litre = ______
13 $\frac{1}{10}$ of 1 km = ______	**14** $\frac{1}{20}$ of 1 kg = ______	**15** $\frac{1}{100}$ of 5 litres = ______
16 $\frac{3}{8}$ of 4000 L = ______	**17** $\frac{1}{3}$ of \$6.60 = ______	**18** $\frac{1}{4}$ of 1 hour = ______
19 $\frac{3}{10}$ of 9 kg = ______	**20** $\frac{4}{5}$ of 2 m = ______	**21** $\frac{9}{11}$ of \$99 = ______
22 $\frac{1}{4}$ of \$120 = ______	**23** $\frac{2}{3}$ of 24 hours = ______	**24** $\frac{3}{5}$ of 45 km = ______
25 $\frac{2}{3}$ of 1 hour = ______	**26** $\frac{5}{6}$ of 3000 m = ______	**27** $\frac{2}{3}$ of \$6.45 = ______
28 $\frac{1}{2}$ of 10 kg = ______	**29** $\frac{1}{3}$ of 60 g = ______	**30** $\frac{3}{7}$ of \$56 = ______
31 $\frac{1}{3}$ of 591 = ______	**32** $\frac{3}{4}$ of 2 L = ______	**33** $\frac{2}{5}$ of 1 hectare = ______

'Of' means 'times' or 'multiplied by'.

UNIT 55 Fractions and Problem Solving

Find the answers to the following problems.

1 Find the average of $\frac{1}{12}$ and $\frac{2}{3}$.

2 Find the square root of $6\frac{1}{4}$.

3 How many times can $\frac{1}{4}$ be subtracted from 6?

4 Find two fractions whose sum is 1 and whose difference is $\frac{1}{2}$.

5 If you multiply a fraction by 25 and add 7, the answer is 17. Find the fraction.

6 Find the sum of $2\frac{1}{2}$, $1\frac{3}{4}$ and 7.

7 By how much does $15\frac{3}{4}$ exceed $8\frac{1}{2}$?

8 How many halves are in 8 apples?

9 How many fifths are in 2 wholes?

10 Five people each work $1\frac{1}{4}$ of an hour overtime. How many hours overtime are worked?

11 I bought 10 bottles of drink. Each bottle contained $\frac{3}{5}$ litre. How many litres of drink did I buy?

12 Kerry had \$24 and spent $\frac{2}{3}$ of it. How much money was left?

13 Matt had to do 60 maths questions for homework but could only do $\frac{3}{5}$ of them. How many were not done?

14 Jean had \$6.60 and spent $\frac{7}{11}$ of it. How much did she spend?

UNIT 56 

Quick Questions

Find the answers to the following.

1 $\frac{3}{2} \times \square = 15$

2 $\frac{3}{7} \times \square = 12$

3 $\square \times 100 = 12$

4 $\frac{3}{5} \times \square = 36$

5 $\frac{3}{7} = \frac{12}{\square}$

6 Write $\frac{53}{9}$ as a mixed number. __________

7 Write $5\frac{3}{7}$ as an improper fraction. __________

8 $\frac{1}{5} < \frac{1}{9}$ True or false? __________

9 $\frac{3}{5} + \frac{1}{5}$ = __________

10 $\frac{4}{15} + \frac{7}{15}$ = __________

11 $\frac{5}{7} + \frac{4}{9}$ = __________

12 $1\frac{1}{2} + 3\frac{1}{4}$ = __________

13 $\frac{11}{17} - \frac{3}{17}$ = __________

14 $6 - \frac{1}{6}$ = __________

15 $\frac{5}{6} - \frac{1}{6}$ = __________

16 $5\frac{1}{2} - 3\frac{1}{3}$ = __________

17 Find $\frac{3}{5}$ of \$60. = __________

18 $\frac{1}{3}$ of $\frac{1}{3}$ = __________

19 $\frac{6}{7} \div \frac{15}{28}$ = __________

20 $\frac{8}{9} \times \frac{5}{6}$ = __________

21 $\left(\frac{2}{3}\right)^2$ = __________

22 $9 \times \frac{3}{4}$ = __________

23 $15 \div \frac{3}{4}$ = __________

24 $16\frac{4}{5} - 12\frac{1}{2}$ = __________

25 $\frac{3}{4}$ of 10 litres = __________

26 Product of 16 and $4\frac{1}{2}$ = __________

27 $5 \div 5\frac{3}{5}$ = __________

28 $\frac{5}{6} \div \left(\frac{3}{10} \div \frac{2}{5}\right)$ = __________

29 The reciprocal of $\frac{9}{4}$ = __________

30 $\sqrt{\frac{4}{9}} \times \sqrt{\frac{25}{36}}$ = __________

UNIT 57 Stay in Touch

A magic square is a square in which all the columns, all the rows and all the diagonals add to give the same total.

Complete the following magic squares.

1

3	$2\frac{1}{2}$	
$1\frac{1}{2}$		
3		

2

$3\frac{1}{6}$		
	$3\frac{2}{3}$	
	$3\frac{1}{3}$	$4\frac{1}{6}$

3

5	$2\frac{1}{2}$	
$1\frac{1}{2}$		
4		

4

$3\frac{3}{4}$	$1\frac{1}{4}$	
$\frac{1}{4}$		$4\frac{1}{4}$
$2\frac{3}{4}$		

5

$1\frac{2}{3}$		
	$2\frac{1}{6}$	
	$1\frac{5}{6}$	$2\frac{2}{3}$

6

$1\frac{1}{2}$		
	$4\frac{1}{2}$	
	$\frac{1}{2}$	$7\frac{1}{2}$

7

$7\frac{1}{2}$		
	$6\frac{1}{2}$	
	$10\frac{1}{2}$	$5\frac{1}{2}$

8

$5\frac{2}{3}$	$\frac{2}{3}$	
$6\frac{2}{3}$		
$1\frac{2}{3}$		

9

$8\frac{1}{2}$		
	$7\frac{1}{2}$	
	$11\frac{1}{2}$	$6\frac{1}{2}$

Simplify the following.

10 $\frac{1}{4} + \frac{1}{2} \times 6$ = ____________________

____

11 $\frac{1}{4} \times 20 + 1\frac{3}{4}$ = ____________________

____

12 $\frac{5}{12} \div \frac{1}{3} + \frac{1}{7}$ = ____________________

____

13 $\frac{5}{7} \times \frac{7}{8} + 2\frac{1}{4}$ = ____________________

____

14 $5\frac{3}{4} - \frac{3}{8}$ of $\frac{8}{15}$ = ____________________

____

15 $5\frac{1}{8} + \frac{1}{4}$ of $\frac{5}{8}$ = ____________________

____

16 $\frac{3}{4}$ of $5\frac{1}{3} - \frac{1}{2}$ of $\frac{6}{7}$ = ____________________

____

17 $\frac{1}{2}\left(\frac{3}{4} + \frac{5}{8}\right)$ = ____________________

____

18 $\left(\frac{5}{6} + \frac{1}{3}\right) \div \frac{14}{3}$ = ____________________

____

19 $\left(\frac{1}{3} - \frac{1}{4}\right) \div \frac{1}{3} + \frac{2}{3}$ = ____________________

____

UNIT 58  Stop. Revise. Check.

☞ When rounding a fraction, if the fraction part is $\geqslant \frac{1}{2}$ round it up, and if the fraction part is $< \frac{1}{2}$ round it down.

e.g. In the fraction $5\frac{6}{7}$, since $\frac{6}{7} > \frac{1}{2}$ round it up. $\therefore\ 5\frac{6}{7} = 6$ to the nearest whole number.

☞ Sometimes the word 'of' is used for multiplication.

e.g. $\frac{1}{4}$ of $20 = \frac{1}{4} \times 20 = 5$

☞ To find the square of a fraction, square the numerator separately and square the denominator separately.

e.g. (a) $\left(\frac{2}{3}\right)^2 = \frac{2^2}{3^2} = \frac{4}{9}$ (b) $\left(2\frac{1}{2}\right)^2 = \left(\frac{5}{2}\right)^2 = \frac{5^2}{2^2} = \frac{25}{4} = 6\frac{1}{4}$

☞ To find the square root of a fraction, take the square root of the numerator separately and the square root of the denominator separately.

e.g. (a) $\sqrt{\frac{4}{9}} = \frac{\sqrt{4}}{\sqrt{9}} = \frac{2}{3}$ (b) $\sqrt{\frac{81}{121}} = \frac{\sqrt{81}}{\sqrt{121}} = \frac{9}{11}$

☞ The BODMAS rule is very helpful in the order of operations. Work in the order given by the word:

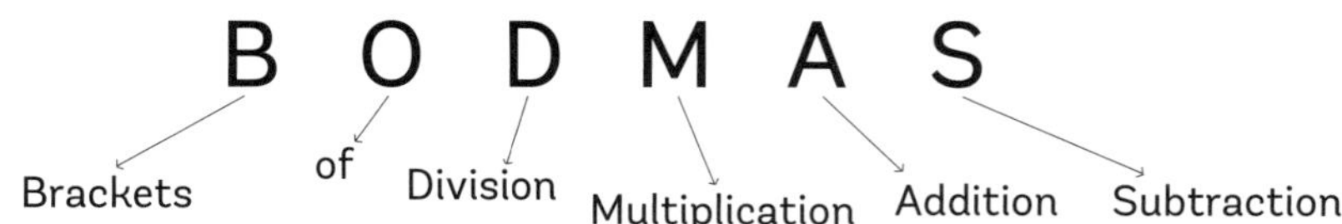

e.g. $\frac{3}{8} + \frac{1}{2} \times \frac{1}{4} = \frac{3}{8} + \frac{1}{8} = \frac{4}{8} = \frac{1}{2}$

☞ The following steps can also be followed in the order of operations:

1. Grouping symbols (brackets) should be calculated first.
2. Multiplication or division should be calculated next, whichever comes first, working from left to right.
3. Addition or subtraction should be calculated next, whichever comes first, working from left to right.

e.g. $\frac{1}{4} \times \frac{1}{3} + \frac{2}{3} \times \frac{1}{2} = \frac{1}{12} + \frac{1}{3} = \frac{1}{12} + \frac{4}{12} = \frac{1+4}{12} = \frac{5}{12}$

UNIT 59 Test on Section 5

Instructions: Do not use a calculator.
Attempt all questions..

Time allowed: 25 minutes **Total marks:** 38

Marks

Question 1: Round to the nearest whole number.

a $\frac{7}{8}$ ______ **b** $3\frac{1}{4}$ ______ **c** $5\frac{3}{8}$ ______ **d** $7\frac{9}{10}$ ______ 4

Question 2: Find the following products.

a $\frac{1}{5}$ of 60 = ______ **b** $\frac{1}{15}$ of 90 = ______ **c** $\frac{3}{4}$ of 48 = ______ **d** $\frac{5}{6}$ of 540 = ______ 4

Question 3: Find the value of the following.

a $\frac{5}{6} \times \frac{5}{6}$ = ______ **b** $1\frac{1}{2} \times 1\frac{1}{2}$ = ______ **c** $\left(\frac{7}{8}\right)^2$ = ______ **d** $\left(3\frac{1}{4}\right)^2$ = ______

______ ______ ______ ______ 4

Question 4: Find the square root of each of the following..

a $\sqrt{\frac{25}{64}}$ = ______ **b** $\sqrt{\frac{36}{81}}$ = ______ **c** $\sqrt{\frac{49}{121}}$ = ______ **d** $\sqrt{\frac{121}{169}}$ = ______

______ ______ ______ ______ 4

Question 5: Use the order of operations to simplify the following.

a $\frac{3}{5} + 1\frac{1}{2} \div \frac{3}{4}$ = ______ **b** $5\frac{2}{9} + 2\frac{1}{2} \div 2\frac{1}{4}$ = ______

c $3\frac{1}{2} + \frac{9}{10}$ of $6\frac{3}{4}$ = ______ **d** $3\frac{7}{10} \div \frac{7}{10} + 2\frac{1}{4}$ = ______ 8

Question 6: Find the value of the following.

a $\frac{3}{4}$ of 64 = ______ **b** $\frac{1}{5}$ of \$120 = ______ **c** $\frac{9}{10}$ of 360 kg = ______

d $\frac{2}{3}$ of 48 hours = ______ **e** $\frac{3}{5}$ of 65 km = ______ **f** $\frac{2}{3}$ of \$81.60 = ______ 6

Question 7: Find the answers to the following problems.

a Find the average of $\frac{3}{5}$ and $1\frac{2}{3}$. ______

b How many thirds are there in 5 wholes? ______

c By how much does $16\frac{5}{6}$ exceed $7\frac{1}{2}$? ______

d Find the square root of $5\frac{1}{16}$. ______  8

Total marks = 38

SECTION 6

UNIT 60 Converting Fractions into Decimals

If the denominator of a fraction has a power of 10 then the number of decimal places is the same as the number of zeros after the 1 in the denominator.

$\frac{7}{10}$ is the same as $7 \div 10$

Example: Change the following fractions into decimals.

(a) $\frac{7}{10}$ (b) $\frac{53}{100}$ (c) $\frac{369}{1000}$ (d) $\frac{58}{1000}$

Solution:

(a) $\frac{7}{10} = 7 \div 10$
$= 0.7$

(b) $\frac{53}{100} = 53 \div 100$
$= 0.53$

(c) $\frac{369}{1000} = 369 \div 1000$
$= 0.369$

(d) $\frac{58}{1000} = 58 \div 1000$
$= 0.058$

Change the following fractions into decimals.

1 $\frac{3}{10}$ = ______	**2** $\frac{1}{10}$ = ______	**3** $\frac{9}{10}$ = ______	**4** $\frac{2}{10}$ = ______
5 $\frac{13}{100}$ = ______	**6** $\frac{29}{100}$ = ______	**7** $\frac{83}{100}$ = ______	**8** $\frac{91}{100}$ = ______
9 $\frac{513}{1000}$ = ______	**10** $\frac{625}{1000}$ = ______	**11** $\frac{334}{1000}$ = ______	**12** $\frac{576}{1000}$ = ______
13 $\frac{12}{10}$ = ______	**14** $\frac{17}{10}$ = ______	**15** $\frac{59}{10}$ = ______	**16** $\frac{83}{10}$ = ______
17 $\frac{127}{100}$ = ______	**18** $\frac{361}{100}$ = ______	**19** $\frac{521}{100}$ = ______	**20** $\frac{963}{100}$ = ______
21 $\frac{4637}{1000}$ = ______	**22** $\frac{8934}{1000}$ = ______	**23** $\frac{6537}{1000}$ = ______	**24** $\frac{9688}{1000}$ = ______
25 $\frac{63}{100}$ = ______	**26** $\frac{59}{100}$ = ______	**27** $\frac{13}{100}$ = ______	**28** $\frac{97}{100}$ = ______
29 $\frac{289}{100}$ = ______	**30** $\frac{273}{100}$ = ______	**31** $\frac{567}{1000}$ = ______	**32** $\frac{936}{1000}$ = ______
33 $\frac{169}{100}$ = ______	**34** $\frac{537}{100}$ = ______	**35** $\frac{689}{1000}$ = ______	**36** $\frac{8}{10}$ = ______
37 $\frac{886}{1000}$ = ______	**38** $\frac{669}{1000}$ = ______	**39** $\frac{561}{1000}$ = ______	**40** $\frac{397}{1000}$ = ______
41 $\frac{443}{100}$ = ______	**42** $\frac{923}{1000}$ = ______	**43** $\frac{369}{1000}$ = ______	**44** $\frac{568}{100}$ = ______

UNIT 61 Converting Fractions into Decimals

Sometimes the denominator is not a power of 10 but is a number that can be easily converted to give a power of 10. This means we convert the fraction to an equivalent fraction with a power of 10 in the denominator and it can then easily be changed into a decimal.

Example: Change the following fractions into decimals.

(a) $\frac{2}{5}$ (b) $\frac{7}{25}$ (c) $\frac{5}{8}$ (d) $\frac{9}{20}$

Solution:

(a) $\frac{2}{5} = \frac{2 \times 2}{5 \times 2} = \frac{4}{10} = 0.4$

(b) $\frac{7}{25} = \frac{7 \times 4}{25 \times 4} = \frac{28}{100} = 0.28$

(c) $\frac{5}{8} = \frac{5 \times 125}{8 \times 125} = \frac{625}{1000} = 0.625$

(d) $\frac{9}{20} = \frac{9 \times 5}{20 \times 5} = \frac{45}{100} = 0.45$

Change the following fractions into decimals.

1 $\frac{3}{5}$ = ______ **2** $\frac{13}{25}$ = ______ **3** $\frac{9}{25}$ = ______

4 $\frac{3}{4}$ = ______ **5** $\frac{11}{20}$ = ______ **6** $\frac{27}{50}$ = ______

7 $\frac{23}{50}$ = ______ **8** $\frac{17}{25}$ = ______ **9** $\frac{3}{8}$ = ______

10 $\frac{3}{40}$ = ______ **11** $\frac{163}{200}$ = ______ **12** $\frac{17}{40}$ = ______

13 $\frac{19}{200}$ = ______ **14** $\frac{19}{20}$ = ______ **15** $\frac{193}{200}$ = ______

16 $\frac{33}{50}$ = ______ **17** $\frac{11}{25}$ = ______ **18** $\frac{69}{200}$ = ______

19 $\frac{469}{500}$ = ______ **20** $\frac{383}{500}$ = ______ **21** $\frac{243}{500}$ = ______

22 $\frac{121}{125}$ = ______ **23** $\frac{39}{50}$ = ______ **24** $\frac{7}{8}$ = ______

25 $\frac{273}{500}$ = ______ **26** $\frac{81}{200}$ = ______ **27** $\frac{121}{200}$ = ______

28 $\frac{41}{50}$ = ______ **29** $\frac{83}{125}$ = ______ **30** $3\frac{7}{20}$ = ______

UNIT 62 Converting Decimals into Fractions

Decimals are also fractions whose denominators are 10 or a power of 10.

For example, $\frac{1}{10}$, $\frac{7}{10}$, $2\frac{9}{10}$, $5\frac{3}{100}$ are all decimal fractions.

To convert a decimal to a fraction, write the denominator by starting with a '1' and adding the same number of zeros after it as there are numerals after the decimal point.

Fractions are in simplest form when the numerator and denominator are as small as possible.

Example: Change the following decimals into fractions.

(a) 0.3 (b) 1.7 (c) 0.21 (d) 0.123

Solution:

(a) 0.3 $= \frac{3}{10}$

(b) 1.7 $= \frac{17}{10} = 1\frac{7}{10}$

(c) 0.21 $= \frac{21}{100}$

(d) 0.123 $= \frac{123}{1000}$

Change the following decimals into fractions.

1 0.1 = ______	**2** 0.3 = ______	**3** 0.7 = ______	**4** 1.9 = ______
5 0.11 = ______	**6** 0.13 = ______	**7** 0.17 = ______	**8** 0.19 = ______
9 0.21 = ______	**10** 0.23 = ______	**11** 0.27 = ______	**12** 0.29 = ______
13 2.31 = ______	**14** 0.33 = ______	**15** 5.37 = ______	**16** 0.39 = ______
17 2.41 = ______	**18** 0.43 = ______	**19** 3.47 = ______	**20** 0.53 = ______
21 0.59 = ______	**22** 7.61 = ______	**23** 0.71 = ______	**24** 0.73 = ______
25 5.79 = ______	**26** 0.83 = ______	**27** 6.89 = ______	**28** 0.91 = ______
29 7.93 = ______	**30** 0.97 = ______	**31** 9.99 = ______	**32** 0.101 = ______
33 0.103 = ______	**34** 2.121 = ______	**35** 5.321 = ______	**36** 8.129 = ______
37 10.519 = ______	**38** 8.213 = ______	**39** 9.353 = ______	**40** 5.729 = ______

UNIT 63 Converting Decimals into Fractions

When we convert a decimal to a fraction, sometimes it needs to be simplified further.

Example: Write the following decimals in simplest fraction form.

(a) 0.2 (b) 0.5 (c) 0.16 (d) 1.25

Solution:

(a) 0.2 $= \frac{2}{10} = \frac{1}{5}$

(b) 0.5 $= \frac{5}{10} = \frac{1}{2}$

(c) 0.16 $= \frac{\overset{4}{\cancel{16}}}{\underset{25}{\cancel{100}}} = \frac{4}{25}$

(d) 1.25 $= \frac{125}{100} = \frac{5}{4} = 1\frac{1}{4}$

Reduce answers to their lowest terms.

Change the following to the simplest fraction form.

1 0.8 = ________	**2** 0.4 = ________	**3** 0.6 = ________
4 0.5 = ________	**5** 0.15 = ________	**6** 0.25 = ________
7 0.05 = ________	**8** 0.16 = ________	**9** 0.24 = ________
10 1.25 = ________	**11** 2.25 = ________	**12** 3.25 = ________
13 0.64 = ________	**14** 0.85 = ________	**15** 0.72 = ________
16 0.68 = ________	**17** 0.65 = ________	**18** 0.75 = ________
19 8.45 = ________	**20** 3.24 = ________	**21** 0.04 = ________
22 0.96 = ________	**23** 0.88 = ________	**24** 1.12 = ________
25 3.5 = ________	**26** 8.4 = ________	**27** 7.2 = ________
28 2.75 = ________	**29** 5.25 = ________	**30** 0.66 = ________
31 3.2 = ________	**32** 5.6 = ________	**33** 7.35 = ________

UNIT 64 Converting Fractions into Percentages

To change a fraction into a percentage, multiply it by 100 and attach a % sign to it.

'Per cent' means 'out of 100'.

Example: Change the following fractions into percentages.

(a) $\frac{9}{10}$ (b) $\frac{3}{5}$ (c) $1\frac{3}{4}$ (d) $\frac{7}{40}$

Solution:

(a) $\frac{9}{10}$
$= \frac{9}{10} \times 100\%$
$= 90\%$

(b) $\frac{3}{5}$
$= \frac{3}{\cancel{5}_1} \times \cancel{100}^{20}\%$
$= 60\%$

(c) $1\frac{3}{4}$
$= \frac{7}{\cancel{4}_1} \times \cancel{100}^{25}\%$
$= 175\%$

(d) $\frac{7}{40}$
$= \frac{7}{40} \times 100\%$
$= \frac{35}{2} = 17\frac{1}{2}\%$

Change the following fractions into percentages.

1 $\frac{1}{2}$	= ________	**2** $\frac{1}{5}$	= ________	**3** $\frac{1}{4}$	= ________
4 $\frac{2}{5}$	= ________	**5** $\frac{3}{4}$	= ________	**6** $\frac{1}{8}$	= ________
7 $\frac{7}{8}$	= ________	**8** $\frac{1}{20}$	= ________	**9** $\frac{1}{50}$	= ________
10 $\frac{3}{20}$	= ________	**11** $1\frac{1}{2}$	= ________	**12** $2\frac{1}{4}$	= ________
13 $\frac{7}{100}$	= ________	**14** $\frac{9}{100}$	= ________	**15** $\frac{11}{100}$	= ________
16 $\frac{17}{50}$	= ________	**17** $\frac{3}{25}$	= ________	**18** $\frac{23}{25}$	= ________
19 $\frac{49}{50}$	= ________	**20** $\frac{3}{10}$	= ________	**21** $\frac{69}{100}$	= ________
22 $\frac{81}{100}$	= ________	**23** $\frac{33}{100}$	= ________	**24** $\frac{83}{200}$	= ________
25 $\frac{189}{200}$	= ________	**26** $\frac{27}{50}$	= ________	**27** $\frac{68}{100}$	= ________
28 $\frac{47}{50}$	= ________	**29** $\frac{9}{20}$	= ________	**30** $\frac{7}{40}$	= ________
31 $\frac{81}{500}$	= ________	**32** $\frac{539}{1000}$	= ________	**33** $\frac{637}{1000}$	= ________

UNIT 65 Converting Percentages into Fractions

'Per cent' means '**out of 100**'. Therefore, to change a percentage into a fraction, divide it by 100.

Example: Change the following percentages into fractions in simplest form.

(a) 33% (b) 75% (c) 85% (d) 200%

Solution:

(a) 33% $= \frac{33}{100}$

(b) 75% $= \frac{75}{100} = \frac{3}{4}$

(c) 85% $= \frac{85}{100} = \frac{17}{20}$

(d) 200% $= \frac{200}{100} = 2$

$\frac{33}{100}$ is the same as $33 \div 100$

Write the following percentages as fractions.

1	13%	________	**2**	29%	________	**3**	69%	________
4	33%	________	**5**	9%	________	**6**	27%	________
7	40%	________	**8**	100%	________	**9**	64%	________
10	10%	________	**11**	80%	________	**12**	50%	________
13	88%	________	**14**	25%	________	**15**	150%	________
16	75%	________	**17**	15%	________	**18**	7%	________
19	17%	________	**20**	96%	________	**21**	225%	________
22	635%	________	**23**	660%	________	**24**	510%	________
25	90%	________	**26**	180%	________	**27**	136%	________
28	270%	________	**29**	512%	________	**30**	360%	________
31	820%	________	**32**	364%	________	**33**	120%	________

UNIT 66 Converting Decimals into Percentages

To change a decimal into a percentage, multiply it by 100 and attach a % sign to it.

When multiplying by 100, move the decimal point two places to the right.

Example: Change the following decimals into percentages.

(a) 0.38 (b) 0.07 (c) 0.431 (d) 1.25

Solution:

(a) $0.38 \times 100 = 38\%$

(b) $0.07 \times 100 = 7\%$

(c) $0.431 \times 100 = 43.1\%$

(d) $1.25 \times 100 = 125\%$

Write the following decimals as percentages.

1 0.75 = ______	**2** 0.50 = ______	**3** 0.25 = ______	**4** 0.23 = ______
5 0.15 = ______	**6** 0.83 = ______	**7** 0.33 = ______	**8** 0.1 = ______
9 0.2 = ______	**10** 0.12 = ______	**11** 0.19 = ______	**12** 0.31 = ______
13 0.8 = ______	**14** 0.125 = ______	**15** 1.45 = ______	**16** 0.3 = ______
17 0.5 = ______	**18** 0.02 = ______	**19** 2.38 = ______	**20** 0.85 = ______
21 0.96 = ______	**22** 0.01 = ______	**23** 0.06 = ______	**24** 0.09 = ______
25 0.28 = ______	**26** 0.55 = ______	**27** 0.67 = ______	**28** 0.4 = ______
29 0.87 = ______	**30** 0.96 = ______	**31** 0.53 = ______	**32** 0.89 = ______
33 0.829 = ______	**34** 0.296 = ______	**35** 0.635 = ______	**36** 0.58 = ______
37 0.61 = ______	**38** 0.215 = ______	**39** 0.39 = ______	**40** 0.036 = ______
41 0.091 = ______	**42** 0.905 = ______	**43** 0.07 = ______	**44** 0.11 = ______
45 0.001 = ______	**46** 0.237 = ______	**47** 0.386 = ______	**48** 0.613 = ______
49 0.252 = ______	**50** 2.3 = ______	**51** 8.94 = ______	**52** 0.167 = ______
53 0.298 = ______	**54** 0.966 = ______	**55** 0.345 = ______	**56** 0.514 = ______
57 0.691 = ______	**58** 3.92 = ______	**59** 8.36 = ______	**60** 2.65 = ______
61 0.697 = ______	**62** 0.357 = ______	**63** 0.123 = ______	**64** 0.321 = ______

UNIT 67 Converting Percentages into Decimals

In order to change a percentage into a decimal, divide it by 100.

Example: Change the following percentages into decimals.

(a) 23% (b) 145% (c) 68.3% (d) 96%

Solution:

(a) 23% $= \frac{23}{100} = 0.23$

(b) 145% $= \frac{145}{100} = 1.45$

(c) 68.3% $= \frac{68.3}{100} = \frac{683}{1000} = 0.683$

(d) 96% $= \frac{96}{100} = 0.96$

100% means 1
50% means 0.5
25% means 0.25

Write the following percentages as decimals.

1 25% = ______	**2** 20% = ______	**3** 10% = ______
4 24% = ______	**5** 53% = ______	**6** 27% = ______
7 60% = ______	**8** 87% = ______	**9** 99% = ______
10 50% = ______	**11** 100% = ______	**12** 75% = ______
13 62.5% = ______	**14** 55% = ______	**15** 78.3% = ______
16 63.5% = ______	**17** 30% = ______	**18** 40% = ______
19 65% = ______	**20** 38.3% = ______	**21** 39.6% = ______
22 92% = ______	**23** 59% = ______	**24** 11.5% = ______
25 63% = ______	**26** 25.3% = ______	**27** 24.9% = ______
28 36.5% = ______	**29** 45.3% = ______	**30** 2.5% = ______
31 68% = ______	**32** 28% = ______	**33** 0.03% = ______
34 15.7% = ______	**35** 73.8% = ______	**36** 96.3% = ______

UNIT 68 Fill in the Missing Percentages, Fractions and Decimals

Fractions, decimals and percentages are closely related to one another and can be converted from one form to another.

Complete the following table by filling in the missing fractions, decimals and percentages.

	Fraction	Decimal	Percentage
1	$\frac{2}{5}$		
2			5%
3		0.85	
4	$\frac{3}{10}$		
5			10%
6		0.125	
7	$\frac{7}{25}$		
8			20%
9	$\frac{4}{5}$		
10		0.4	
11			25%
12	$\frac{9}{20}$		
13		0.8	

	Fraction	Decimal	Percentage
14			50%
15		0.72	
16	$\frac{3}{8}$		
17			75%
18		0.55	
19	$\frac{19}{20}$		
20			100%
21		0.85	
22	$\frac{1}{8}$		
23			120%
24		0.65	
25	$1\frac{3}{4}$		
26			150%

UNIT 69  Quick Questions

Change the following fractions into decimals.

1 $\frac{1}{10}$ ________ 2 $\frac{2}{10}$ ________ 3 $\frac{3}{10}$ ________ 4 $\frac{7}{100}$ ________

5 $\frac{2}{5}$ ________ 6 $\frac{3}{25}$ ________ 7 $\frac{9}{20}$ ________ 8 $\frac{3}{50}$ ________

Change the following decimals into fractions.

9 0.5 = ________ 10 0.7 = ________ 11 0.9 = ________ 12 0.125 = ________

13 0.75 = ________ 14 1.5 = ________ 15 3.3 = ________ 16 7.8 = ________

Change the following fractions into percentages.

17 $\frac{9}{10}$ ________ 18 $\frac{3}{25}$ ________ 19 $\frac{7}{100}$ ________ 20 $\frac{3}{4}$ ________

21 $\frac{83}{100}$ ________ 22 $\frac{67}{200}$ ________ 23 $\frac{1}{4}$ ________ 24 $\frac{1}{2}$ ________

Change the following percentages into fractions.

25 11% = ________ 26 23% = ________ 27 29% = ________ 28 80% = ________

29 10% = ________ 30 75% = ________ 31 120% = ________ 32 225% = ________

Change the following decimals into percentages.

33 0.25 = ________ 34 0.37 = ________ 35 0.81 = ________ 36 0.93 = ________

37 0.07 = ________ 38 2.5 = ________ 39 0.55 = ________ 40 3.65 = ________

Change the following percentages into decimals.

41 10% = ________ 42 20% = ________ 43 35% = ________ 44 69% = ________

45 120% = ________ 46 12.5% = ________ 47 63% = ________ 48 58% = ________

49 28.3% = ________ 50 85% = ________ 51 40% = ________ 52 60% = ________

UNIT 70  Stay in Touch

Find the answers to the following.

1 $\frac{1}{5} \times \square = 10$

2 $\square \times 120 = 30$

3 $\frac{3}{7} \times \square = 36$

4 $\frac{3}{4} = \frac{15}{\square}$

5 $\frac{3}{5} \times \square = 60$

6 Write $3\frac{1}{3}$ as an improper fraction. ____________

7 Write $\frac{39}{11}$ as a mixed number. ____________

8 $\frac{1}{2} < \frac{1}{7}$ True or false? ____________

9 $\frac{2}{7} + \frac{1}{7}$ = ____________

10 $\frac{7}{15} + \frac{2}{15}$ = ____________

11 $\frac{5}{7} + \frac{2}{9}$ = ____________

12 $\frac{1}{4} + 5\frac{1}{4}$ = ____________

13 $\frac{13}{17} - \frac{2}{17}$ = ____________

14 $7 - \frac{1}{7}$ = ____________

15 $\frac{3}{4} - \frac{5}{8}$ = ____________

16 $5\frac{3}{4} - 3\frac{1}{3}$ = ____________

17 Find $\frac{3}{8}$ of \$640. = ____________

18 $\frac{1}{3}$ of $\frac{1}{3}$ = ____________

19 $\frac{7}{8} \times \frac{12}{15}$ = ____________

20 $\frac{5}{7} \div \frac{15}{49}$ = ____________

21 $\left(\frac{3}{5}\right)^2 \div \frac{66}{125}$ = ____________

22 $2\frac{1}{2} + 2\frac{3}{4} + 7\frac{1}{2}$ = ____________

23 Product of 18 and $4\frac{1}{2}$ = ____________

24 Find $\frac{2}{5}$ of 80 litres. ____________

25 $9\frac{11}{12} \times 0 \times \frac{5}{6}$ = ____________

26 $5 \div 5\frac{3}{5}$ = ____________

27 $\frac{9}{10} \times \left(\frac{3}{5} + \frac{2}{3}\right)$ = ____________

28 $\frac{2}{5}$ of $\frac{2}{5}$ = ____________

29 Write $\frac{200}{9}$ as a mixed number. ____________

30 Express 35% as a fraction. ____________

UNIT 71

Stop. Revise. Check.

- Fractions, decimals and percentages are closely related to one another and can be easily converted from one to another.

  e.g. $\frac{1}{4} = 0.25 = 25\%$

- If the denominator of a fraction has a power of 10 then the number of decimal places is the same as the number of zeros after the 1 in the denominator.

  e.g. (a) $\frac{9}{10} = 9 \div 10 = 0.9$ (b) $\frac{67}{100} = 67 \div 100 = 0.67$

- If the denominator is not a power of 10 but is a number that can be easily converted to give a power of 10, simply convert the fraction to a fraction with a power of 10 in the denominator before changing it into a decimal.

  e.g. (a) $\frac{3}{5} = \frac{3 \times 2}{5 \times 2} = \frac{6}{10} = 6 \div 10 = 0.6$ (b) $\frac{19}{20} = \frac{19 \times 5}{20 \times 5} = \frac{95}{100} = 95 \div 100 = 0.95$

- To change a decimal into a fraction, write the denominator by starting with a '1' and adding the same number of zeros as there are numerals after the decimal point.

  e.g. (a) $0.7 = \frac{7}{10}$ (b) $0.37 = \frac{37}{100}$ (c) $3.49 = \frac{349}{100} = 3\frac{49}{100}$

- To change a fraction or a decimal into a percentage, multiply by 100 and attach a % sign.

  Example: Change the following into percentages.

  (a) $\frac{3}{10}$ (b) $\frac{2}{5}$ (c) 1.25 (d) 0.32

  Solution:

  (a) $\frac{3}{10} \times 100 = 30\%$ (b) $\frac{2}{5} \times 100 = 40\%$ (c) $1.25 \times 100 = 125\%$ (d) $0.32 \times 100 = 32\%$

- 'Per cent' means 'out of 100'.

  e.g. (a) $23\% = \frac{23}{100}$ (b) $39\% = \frac{39}{100}$

- To change a percentage into a fraction or a decimal, simply divide it by 100.

  e.g. (a) $33\% = \frac{33}{100}$ (b) $25\% = \frac{25}{100} = 0.25$

  (c) $750\% = \frac{750}{100} = 7\frac{1}{2}$ (d) $38\% = \frac{38}{100} = 0.38$

UNIT 72 Test on Section 6

Instructions: Do not use a calculator.
Attempt all questions.

Time allowed: 25 minutes **Total marks:** 36

Marks

Question 1: Change the following fractions into decimals.

a $\frac{5}{10}$ = ______ **b** $\frac{9}{100}$ = ______ **c** $\frac{23}{100}$ = ______

d $\frac{6}{25}$ = ______ **e** $\frac{7}{20}$ = ______ **f** $\frac{123}{10}$ = ______ 6

Question 2: Change the following decimals into fractions.

a 0.9 = ______ **b** 0.13 = ______ **c** 0.125 = ______

d 1.2 = ______ **e** 30.5 = ______ **f** 2.36 = ______ 6

Question 3: Change the following fractions into percentages.

a $\frac{11}{20}$ = ______ **b** $\frac{16}{25}$ = ______ **c** $\frac{9}{100}$ = ______

d $\frac{3}{4}$ = ______ **e** $\frac{1}{8}$ = ______ **f** $1\frac{1}{4}$ = ______ 6

Question 4: Change the following perentages into fractions.

a 29% = ______ **b** 60% = ______ **c** 125% = ______

d 75% = ______ **e** $62\frac{1}{2}\%$ = ______ **f** $5\frac{1}{4}\%$ = ______ 6

Question 5: Change the following decimals into percentages.

a 0.73 = ______ **b** 8.9 = ______ **c** 6.32 = ______

d 5.4 = ______ **e** 0.25 = ______ **f** 0.625 = ______ 6

Question 6: Change the following percentages into decimals.

a 80% = ______ **b** 60% = ______ **c** 55% = ______

d 180% = ______ **e** 95% = ______ **f** 33% = ______ 6

Total marks = / 36

SECTION 7 EXAM PAPER 1

Instructions: Attempt all questions.
Do **not** use a calculator.
All necessary working must be shown in every question.

Time allowed: 1 hour **Total marks:** 100

Part A Show all necessary working.

Marks

Question 1: Change these mixed numbers into improper fractions.

a $2\frac{1}{3}$ = ________ **b** $9\frac{1}{2}$ = ________ **c** $3\frac{4}{5}$ = ________ **d** $7\frac{1}{10}$ = ________ 4

Question 2: Put in <, = or > to make these true statements.

a $\frac{1}{3}$ ☐ $\frac{1}{9}$ **b** $\frac{5}{8}$ ☐ $\frac{3}{5}$ **c** $\frac{1}{4}$ ☐ $\frac{1}{6}$ **d** $\frac{3}{5}$ ☐ $\frac{7}{8}$ 4

Question 3: State true (T) or false (F) for each of the following.

a $\frac{1}{2} > \frac{1}{4}$ ________ **b** $\frac{1}{7} > \frac{1}{6}$ ________ **c** $\frac{9}{12} = \frac{3}{4}$ ________ **d** $\frac{7}{9} < \frac{11}{18}$ ________ 4

Question 4: Reduce the following to their simplest form.

a $\frac{5}{10}$ = ________ **b** $\frac{36}{48}$ = ________ **c** $\frac{64}{120}$ = ________ **d** $\frac{56}{72}$ = ________ 4

Question 5: Change these improper fractions into mixed numbers.

a $\frac{19}{3}$ = ________ **b** $\frac{70}{9}$ = ________ **c** $\frac{48}{20}$ = ________ **d** $\frac{100}{7}$ = ________ 4

Question 6: Complete the following equivalent fractions.

a $\frac{3}{5} = \frac{18}{\square}$ **b** $\frac{2}{7} = \frac{16}{\square}$ **c** $\frac{5}{6} = \frac{\square}{54}$ **d** $\frac{3}{11} = \frac{\square}{55}$ 4

Question 7: Arrange each set of fractions in ascending order.

a $\frac{1}{3}, \frac{1}{7}, \frac{1}{5}$ ________ **b** $\frac{2}{3}, \frac{3}{5}, \frac{2}{5}$ ________ 2

c $\frac{5}{8}, \frac{1}{2}, \frac{1}{4}$ ________ **d** $\frac{7}{12}, \frac{3}{8}, \frac{1}{3}$ ________ 2

e $\frac{1}{4}, \frac{1}{3}, \frac{1}{6}$ ________ **f** $\frac{1}{2}, \frac{2}{3}, \frac{3}{4}$ ________ 2

Total marks for Part A  30

EXAM PAPER 1 Part B Show all necessary working.

Marks

Question 8: Add and simplify the following fractions.

a $\frac{9}{16}+\frac{9}{16}$ = ________ ________

b $\frac{2}{5}+\frac{2}{7}$ = ________ ________

c $4\frac{3}{5}+2\frac{1}{3}$ = ________ ________

3

Question 9: Subtract and simplify the following fractions.

a $\frac{7}{8}-\frac{5}{8}$ = ________ ________

b $\frac{7}{8}-\frac{1}{2}$ = ________ ________

c $3\frac{1}{6}-1\frac{5}{12}$ = ________ ________

3

Question 10: Multiply and simplify the following fractions.

a $\frac{3}{8}\times\frac{7}{12}$ = ________ ________

b $\frac{10}{27}\times\frac{9}{40}$ = ________ ________

c $5\frac{1}{4}\times 3\frac{1}{3}$ = ________ ________

3

Question 11: Divide and simplify the following fractions.

a $7\div\frac{2}{3}$ = ________ ________

b $\frac{5}{8}\div\frac{16}{25}$ = ________ ________

c $1\frac{5}{9}\div 2\frac{2}{3}$ = ________ ________

3

Question 12: Find the following.

a $\frac{2}{5}$ of 1 m ________ b $\frac{3}{4}$ of 1 hour ________ c $\frac{4}{5}$ of \$100 ________ d $\frac{9}{10}$ of 180º ________ 4

What fraction is:

e 20 cm of 2 m? ____ f 40c of \$5? ____ g 300 g of 3 kg? ____ h 1.6 km of 2 km? ____ 4

Question 13: Simplify the following.

a $\left(\frac{5}{6}+\frac{1}{2}\right)\div\frac{1}{12}$ = ________________

b $3\frac{7}{10}\div\frac{9}{10}+2\frac{1}{4}$ = ________________ 6

c $\frac{5}{6}+2\frac{1}{3}\div\frac{1}{2}$ = ________________

d $\dfrac{\frac{1}{2}+\frac{1}{3}}{\frac{1}{2}-\frac{1}{3}}$ = ________________ 6

Question 14:

a Robert earned \$65 a week and saved $\frac{2}{5}$ of it. How much did he save a week? ________________ 2

b If a bottle of milk contains $\frac{3}{4}$ litre of milk and Shelley drinks $\frac{3}{8}$ of it, how much is left? ________________ 2

c A car travelled 180 km in $2\frac{1}{4}$ hours. Find its speed in kilometres per hour. ________________ 2

d What fraction of a year is a fortnight? ________________ 2

Total marks for Part B 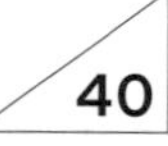 40

EXAM PAPER 1 Part C Show all necessary working.

Marks

Question 15: Find the following.

a $\frac{3}{4}\times\frac{3}{4}$ = ______ **b** $\frac{5}{8}\times\frac{5}{8}$ = ______ **c** $\left(\frac{4}{9}\right)^2$ = ______ **d** $\left(\frac{7}{10}\right)^2$ = ______

______ ______ ______ ______ 4

Question 16: Find the value of the following.

a $\sqrt{\frac{25}{36}}$ = ______ **b** $\sqrt{\frac{64}{121}}$ = ______ **c** $\sqrt{\frac{81}{144}}$ = ______ **d** $\sqrt{\frac{36}{100}}$ = ______

______ ______ ______ ______ 4

Question 17: Change the following fractions into decimals.

a $\frac{3}{10}$ = ______ **b** $\frac{9}{100}$ = ______ **c** $\frac{3}{5}$ = ______ **d** $\frac{9}{20}$ = ______ 4

Question 18: Change the following decimals into fractions.

a 0.61 = ______ **b** 0.12 = ______ **c** 1.25 = ______ **d** 3.6 = ______ 4

Question 19: Change the following fractions into percentages.

a $\frac{8}{10}$ = ______ **b** $\frac{19}{25}$ = ______ **c** $\frac{17}{40}$ = ______ **d** $5\frac{11}{20}$ = ______ 4

Question 20: Change the following percentages into fractions.

a 99% = ______ **b** 64% = ______ **c** 125% = ______ **d** 30% = ______ 4

Question 21:

Complete the following table by filling in the missing fractions, decimals and percentages.

	Fraction	Decimal	Percentage	Marks
a	$\frac{7}{100}$			2
b		0.36		2
c			33%	2

End of Paper

Total marks for Part C / 30

EXAM PAPER 2

Instructions: Attempt all questions.
Do **not** use a calculator.
All necessary working must be shown in every question.

Time allowed: 1 hour **Total marks:** 100

Part A Show all necessary working.

Marks

Question 1: Change these mixed numbers into improper fractions.

a $2\frac{3}{4}$ = ________ b $5\frac{1}{3}$ = ________ c $3\frac{2}{5}$ = ________ d $7\frac{1}{4}$ = ________ 4

Question 2: Change these improper fractions into mixed numbers.

a $\frac{9}{2}$ = ________ b $\frac{13}{4}$ = ________ c $\frac{45}{7}$ = ________ d $\frac{38}{4}$ = ________ 4

Question 3: Reduce the following fractions to their simplest form.

a $\frac{25}{35}$ = ________ b $\frac{50}{70}$ = ________ c $\frac{48}{72}$ = ________ d $\frac{30}{8}$ = ________ 4

Question 4: Complete the following equivalent fractions.

a $\frac{4}{9} = \frac{\square}{45}$ b $\frac{7}{9} = \frac{\square}{63}$ c $\frac{4}{5} = \frac{28}{\square}$ d $\frac{2}{3} = \frac{40}{\square}$ 4

Question 5: Put in $<$, $=$ or $>$ to make these true statements.

a $\frac{1}{5} \square \frac{1}{4}$ b $\frac{2}{3} \square \frac{4}{5}$ c $\frac{9}{10} \square \frac{8}{9}$ d $\frac{54}{90} \square \frac{4}{5}$ 4

Question 6: State whether each of the following is true (T) or false (F).

a $\frac{3}{4} > \frac{3}{7}$ ________ b $\frac{2}{9} < \frac{2}{3}$ ________ c $\frac{5}{12} = \frac{15}{36}$ ________ d $\frac{5}{10} > \frac{4}{6}$ ________ 4

Question 7: Arrange the following in descending order.

a $\frac{1}{2}, \frac{1}{5}, \frac{1}{9}$ ________ b $\frac{3}{5}, \frac{2}{3}, \frac{3}{4}$ ________ 2

c $\frac{3}{7}, \frac{1}{4}, \frac{3}{4}$ ________ d $\frac{5}{8}, \frac{2}{7}, \frac{2}{3}$ ________ 2

e $\frac{5}{9}, \frac{3}{7}, \frac{4}{11}$ ________ f $\frac{2}{5}, \frac{3}{9}, \frac{2}{7}$ ________ 2

Total marks for Part A / 30

EXAM PAPER 2 Part B

Show all necessary working.

Marks

Question 8: Add and simplify the following fractions.

a $\frac{9}{25}+\frac{6}{25}$ = ________ ________ **b** $\frac{5}{7}+\frac{3}{4}$ = ________ ________ **c** $5\frac{1}{4}+6\frac{2}{3}$ = ________ ________ 3

Question 9: Subtract and simplify the following fractions.

a $\frac{9}{17}-\frac{5}{17}$ = ________ ________ **b** $\frac{12}{25}-\frac{8}{25}$ = ________ ________ **c** $8\frac{2}{9}-3\frac{5}{6}$ = ________ ________ 3

Question 10: Multiply the following.

a $\frac{5}{7}\times\frac{3}{8}$ = ________ ________ **b** $\frac{24}{45}\times\frac{15}{36}$ = ________ ________ **c** $8\frac{2}{5}\times 9\frac{3}{10}$ = ________ ________ 3

Question 11: Simplify the following.

a $\frac{18}{55}\div\frac{3}{11}$ = ________ ________ **b** $\frac{12}{39}\div\frac{8}{13}$ = ________ ________ **c** $8\frac{3}{4}\div 2\frac{5}{8}$ = ________ ________ 3

Question 12: Find the following.

a $\frac{5}{6}$ of 300 m ______ **b** $\frac{2}{3}$ of \$6.45 ______ **c** $\frac{1}{4}$ of 64 kg ______ **d** $\frac{1}{3}$ of 591 ______ 4

What fraction is:

e 50 c of \$2? ______ **f** $\frac{2}{5}$ of 800 L? ______ **g** 5 min of 1 h? ______ **h** 60 m of 100 m? ______ 4

Question 13: Simplify the following.

a $\left(\frac{1}{5}+\frac{1}{6}\right)\div\frac{5}{6}+3$ = ________ 6

b $\frac{8}{9}\times 1\frac{1}{3}+2\frac{1}{4}$ = ________

c $5\frac{3}{8}+2\frac{1}{9}-3\frac{2}{5}$ = ________ 6

d $\dfrac{7}{\frac{1}{2}-\frac{1}{3}}$ = ________

Question 14:

a A class has 18 girls and 12 boys. What fraction of the class is girls? ________ ________ 2

b How many times can $\frac{1}{5}$ be subtracted from 6? ________ 2

c $\frac{7}{10}$ of a container is emptied and 6 litres remain. What is the capacity of the container? ________ ________ 2

d The sum of two numbers is 8. One of the numbers is $3\frac{5}{6}$. Find the other number. ________ ________ 2

Total marks for Part B /40

EXAM PAPER 2 Part C

Show all necessary working.

Marks

Question 15: Find the following.

a $\frac{5}{9} \times \frac{5}{9}$ = ________ **b** $\frac{3}{7} \times \frac{3}{7}$ = ________ **c** $\left(\frac{4}{11}\right)^2$ = ________ **d** $\left(\frac{9}{13}\right)^2$ = ________

________ ________ ________ ________ 4

Question 16: Find the value of the following.

a $\sqrt{\frac{25}{81}}$ = ________ **b** $\sqrt{\frac{49}{169}}$ = ________ **c** $\sqrt{\frac{100}{144}}$ = ________ **d** $\sqrt{\frac{9}{64}}$ = ________

________ ________ ________ ________ 4

Question 17: Change the following fractions into decimals.

a $\frac{7}{10}$ = ________ **b** $\frac{11}{100}$ = ________ **c** $\frac{9}{25}$ = ________ **d** $\frac{13}{50}$ = ________ 4

Question 18: Change the following decimals into fractions.

a 0.23 = ________ **b** 0.72 = ________ **c** 2.75 = ________ **d** 5.6 = ________ 4

Question 19: Change the following fractions into percentages.

a $\frac{1}{10}$ = ________ **b** $\frac{17}{20}$ = ________ **c** $\frac{29}{120}$ = ________ **d** $4\frac{1}{2}$ = ________ 4

Question 20: Change the following percentages into fractions.

a 97% = ________ **b** 86% = ________ **c** 33% = ________ **d** 6% = ________ 4

Question 21:

Complete the following table by filling in the missing fractions, decimals and percentages.

	Fraction	Decimal	Percentage
a	$\frac{1}{4}$		
b		0.125	
c			70%

Marks: a 2, b 2, c 2

End of Paper

Total marks for Part C /30

EXAM PAPER 3

Instructions: Attempt all questions.
Do **not** use a calculator.
All necessary working must be shown in every question.

Time allowed: 1 hour **Total marks:** 100

Part A Show all necessary working.

Marks

Question 1: Change these mixed numbers into improper fractions.

a $4\frac{1}{2}$ = ________ **b** $5\frac{3}{8}$ = ________ **c** $9\frac{7}{9}$ = ________ **d** $3\frac{7}{8}$ = ________ 4

Question 2: Change these improper fractions into mixed numbers.

a $\frac{120}{11}$ = ________ **b** $\frac{200}{9}$ = ________ **c** $\frac{96}{7}$ = ________ **d** $\frac{169}{10}$ = ________ 4

Question 3: Put in $<$, $=$ or $>$ to make these true statements.

a $\frac{9}{10}$ ☐ $\frac{5}{7}$ **b** $\frac{51}{63}$ ☐ $\frac{30}{35}$ **c** $\frac{43}{70}$ ☐ $\frac{5}{7}$ **d** $\frac{15}{20}$ ☐ $\frac{3}{4}$ 4

Question 4: State whether each of the following is true (T) or false (F).

a $\frac{9}{15} > \frac{3}{5}$ ________ **b** $\frac{8}{17} < \frac{2}{34}$ ________ **c** $\frac{3}{12} = \frac{15}{60}$ ________ **d** $\frac{3}{7} > \frac{5}{9}$ ________ 4

Question 5: Reduce the following to their simplest form.

a $\frac{63}{99}$ = ________ **b** $\frac{25}{625}$ = ________ **c** $\frac{81}{108}$ = ________ **d** $\frac{96}{144}$ = ________ 4

Question 6: Complete the following equivalent fractions.

a $\frac{5}{12} = \frac{60}{\square}$ **b** $\frac{3}{13} = \frac{27}{\square}$ **c** $\frac{15}{22} = \frac{\square}{66}$ **d** $\frac{8}{15} = \frac{\square}{225}$ 4

Question 7: Arrange each set of fractions in ascending order.

a $\frac{5}{7}, \frac{1}{9}, \frac{2}{3}$ ________________ **b** $\frac{8}{11}, \frac{5}{12}, \frac{3}{4}$ ________________ 2

c $\frac{9}{11}, \frac{12}{15}, \frac{8}{20}$ ________________ **d** $\frac{10}{25}, \frac{3}{18}, \frac{1}{7}$ ________________ 2

e $\frac{16}{17}, \frac{5}{17}, \frac{23}{34}$ ________________ **f** $\frac{8}{14}, \frac{5}{18}, \frac{3}{7}$ ________________ 2

Total marks for Part A ☐ / 30

EXAM PAPER 3 Part B Show all necessary working.

Marks

Question 8: Add and simplify the following fractions.

a $\frac{3}{10}+\frac{1}{10}$ = ________ ________ **b** $\frac{3}{5}+\frac{7}{11}$ = ________ ________ **c** $5\frac{5}{6}+6\frac{3}{4}$ = ________ ________ 3

Question 9: Subtract the following.

a $\frac{9}{14}-\frac{5}{14}$ = ________ ________ **b** $\frac{8}{35}-\frac{2}{5}$ = ________ ________ **c** $5\frac{7}{9}-3\frac{2}{3}$ = ________ ________ 3

Question 10: Multiply the following.

a $\frac{9}{16}\times\frac{24}{27}$ = ________ ________ **b** $\frac{12}{70}\times\frac{14}{15}$ = ________ ________ **c** $8\frac{1}{4}\times2\frac{1}{2}$ = ________ ________ 3

Question 11: Simplify the following.

a $\frac{2}{15}\div\frac{3}{7}$ = ________ ________ **b** $\frac{15}{24}\div\frac{30}{48}$ = ________ ________ **c** $6\frac{2}{3}\div1\frac{1}{3}$ = ________ ________ 3

Question 12: Find the following.

a $\frac{3}{4}$ of 4 h ______ **b** $\frac{5}{8}$ of 6 km ______ **c** $\frac{4}{9}$ of $1800 ______ **d** $\frac{4}{9}$ of 72 ______ 4

What fraction is:

e 15 cm of 2 m?____ **f** 300 mL of 6 L?____ **g** 8 m of 72 m?____ **h** 40 g of 2 kg?____ 4

Question 13: Simplify the following.

a $\left(\frac{2}{5}+\frac{3}{10}\right)\times\left(\frac{1}{8}+\frac{3}{8}\right)$ = ________ **b** $\frac{8}{9}+2\frac{1}{3}\div\frac{3}{4}$ = ________ 6

c $5\frac{1}{3}\div2\frac{2}{3}+3\frac{1}{4}$ = ________ **d** $\dfrac{2\frac{1}{8}}{\frac{5}{6}+\frac{1}{3}}$ = ________ 6

Question 14:

a There are 40 weeks in a school year. At the end of week 16, what fraction of the year is over? ________ 2

b Find the difference between $15\frac{3}{4}$ and $8\frac{1}{2}$. ________ 2

c Find the sum of $8\frac{2}{3}$, $5\frac{3}{4}$ and $2\frac{1}{2}$. ________ 2

d A man spends $250 per week on food. If this is $\frac{3}{4}$ of his weekly wage, what is his weekly wage? ________ 2

Total marks for Part B /40

EXAM PAPER 3 Part C Show all necessary working.

Marks

Question 15: Find the following.

a $\frac{7}{9} \times \frac{7}{9}$ = ________ **b** $\frac{6}{11} \times \frac{6}{11}$ = ________ **c** $\left(\frac{3}{8}\right)^2$ = ________ **d** $\left(\frac{9}{11}\right)^2$ = ________

________ ________ ________ ________ 4

Question 16: Find the value of the following.

a $\sqrt{\frac{9}{16}}$ = ________ **b** $\sqrt{\frac{36}{49}}$ = ________ **c** $\sqrt{\frac{49}{64}}$ = ________ **d** $\sqrt{\frac{81}{121}}$ = ________

________ ________ ________ ________ 4

Question 17: Change the following fractions into decimals.

a $\frac{7}{100}$ = ________ **b** $\frac{25}{200}$ = ________ **c** $\frac{8}{1000}$ = ________ **d** $\frac{175}{500}$ = ________ 4

Question 18: Change the following decimals into fractions.

a 0.67 = ________ **b** 0.15 = ________ **c** 3.75 = ________ **d** 8.5 = ________ 4

Question 19: Change the following fractions into percentages.

a $\frac{50}{100}$ = ________ **b** $\frac{45}{100}$ = ________ **c** $\frac{68}{100}$ = ________ **d** $3\frac{1}{2}$ = ________ 4

Question 20: Change the following percentages into fractions.

a 72% = ________ **b** 37% = ________ **c** 125%= ________ **d** 550%= ________ 4

Question 21:

Complete the following table by filling in the missing fractions, decimals and percentages.

	Fraction	Decimal	Percentage	Marks
a	$\frac{4}{5}$			2
b		0.65		2
c			88%	2

End of Paper

Total marks for Part C 30

EXAM PAPER 4

Instructions: Attempt all questions.
Do **not** use a calculator.
All necessary working must be shown in every question.

Time allowed: 1 hour **Total marks:** 100

Marks

Part A Show all necessary working.

Question 1: Change these mixed numbers into improper fractions.

a $15\frac{1}{2}$ = ______ b $12\frac{3}{4}$ = ______ c $8\frac{7}{12}$ = ______ d $10\frac{5}{6}$ = ______ 4

Question 2: Change these improper fractions into mixed numbers.

a $\frac{87}{9}$ = ______ b $\frac{33}{4}$ = ______ c $\frac{59}{7}$ = ______ d $\frac{123}{5}$ = ______ 4

Question 3: Put in <, = or > to make these true statements.

a $\frac{9}{11}$ ☐ $\frac{5}{6}$ b $\frac{15}{22}$ ☐ $\frac{5}{11}$ c $\frac{8}{9}$ ☐ $\frac{32}{36}$ d $1\frac{2}{5}$ ☐ $1\frac{3}{7}$ 4

Question 4: State whether each of the following is true (T) or false (F).

a $\frac{8}{9} < \frac{15}{16}$ ______ b $\frac{12}{33} > \frac{5}{9}$ ______ c $\frac{7}{12} = \frac{35}{60}$ ______ d $\frac{19}{32} < \frac{15}{33}$ ______ 4

Question 5: Reduce the following fractions to their simplest form.

a $\frac{80}{90}$ = ______ b $\frac{55}{88}$ = ______ c $\frac{96}{132}$ = ______ d $\frac{64}{240}$ = ______ 4

Question 6: Complete the following equivalent fractions.

a $\frac{15}{27} = \frac{\square}{9}$ b $\frac{8}{25} = \frac{\square}{225}$ c $\frac{5}{7} = \frac{65}{\square}$ d $\frac{3}{14} = \frac{27}{\square}$ 4

Question 7: Arrange the following in descending order.

a $\frac{5}{6}, \frac{2}{3}, \frac{8}{9}$ ______ b $\frac{3}{7}, \frac{5}{14}, \frac{7}{21}$ ______ 2

c $\frac{9}{8}, \frac{8}{9}, \frac{8}{19}$ ______ d $\frac{4}{3}, \frac{5}{7}, \frac{9}{12}$ ______ 2

e $\frac{35}{40}, \frac{60}{65}, \frac{38}{76}$ ______ f $\frac{23}{25}, \frac{28}{29}, \frac{11}{20}$ ______ 2

Total marks for Part A /30

EXAM PAPER 4 Part B Show all necessary working.

Marks

Question 8: Add the following fractions.

a $\frac{8}{15}+\frac{3}{15}$ = ________ ________

b $\frac{5}{6}+\frac{3}{8}$ = ________ ________

c $5\frac{7}{8}+3\frac{2}{3}$ = ________ ________ 3

Question 9: Subtract the following.

a $\frac{9}{14}-\frac{6}{14}$ = ________ ________

b $\frac{5}{9}-\frac{1}{3}$ = ________ ________

c $8\frac{1}{4}-5\frac{5}{8}$ = ________ ________ 3

Question 10: Multiply the following.

a $\frac{5}{12}\times\frac{7}{3}$ = ________ ________

b $\frac{18}{25}\times\frac{45}{81}$ = ________ ________

c $6\frac{3}{4}\times 2\frac{1}{4}$ = ________ ________ 3

Question 11: Divide the following.

a $9\frac{1}{2}\div 5\frac{1}{4}$ = ________ ________

b $\frac{16}{36}\div\frac{8}{45}$ = ________ ________

c $7\frac{2}{5}\div 1\frac{1}{5}$ = ________ ________ 3

Question 12: Find the following.

a $\frac{4}{5}$ of 65 ________ **b** $\frac{1}{2}$ of 38 min ________ **c** $\frac{5}{7}$ of \$35 ________ **d** $\frac{6}{11}$ of 550 ________ 4

What fraction is:

e 2 cm of 34 cm?____ **f** 80c of \$16?____ **g** 200 g of 6 kg?____ **h** 4.2 km of 7 km?____ 4

Question 13: Simplify the following.

a $5\frac{5}{8}+2\frac{1}{9}-3\frac{3}{4}$ = ________________ 6

b $3\frac{1}{2}\div 2\frac{2}{3}+1\frac{5}{6}$ = ________________

c $\frac{1}{2}\left(\frac{1}{4}+\frac{1}{8}\right)\div 5\frac{1}{2}$ = ________________ 6

d $\dfrac{\frac{8}{9}}{\frac{1}{2}\times 2\frac{1}{4}\div\frac{1}{8}}$ = ________________

Question 14:

a In a class of 20 students, $\frac{2}{5}$ of the class have blue eyes. How many students have blue eyes? ________________ 

b How many sixths are there in 5? ________________ 

c If Clare takes $\frac{2}{3}$ of an hour to write a page, how long does she take to write 9 pages? ________________ 2

d If Emma walks at a rate of $6\frac{1}{4}$ km per hour, how long would she take to walk 18 km? ________________ 2

Total marks for Part B /40

EXAM PAPER 4 Part C

Show all necessary working.

Marks

Question 15: Find the following.

a $\frac{12}{25} \times \frac{12}{25} =$ ________ **b** $\frac{20}{30} \times \frac{20}{30} =$ ________ **c** $\left(\frac{5}{13}\right)^2 =$ ________ **d** $\left(6\frac{1}{4}\right)^2 =$ ________

________ ________ ________ ________ 4

Question 16: Find the value of the following.

a $\sqrt{\frac{4}{81}} =$ ________ **b** $\sqrt{\frac{9}{169}} =$ ________ **c** $\sqrt{\frac{25}{121}} =$ ________ **d** $\sqrt{2\frac{7}{9}} =$ ________

________ ________ ________ ________ 4

Question 17: Change the following fractions into decimals.

a $\frac{5}{10} =$ ________ **b** $\frac{3}{4} =$ ________ **c** $\frac{7}{25} =$ ________ **d** $\frac{125}{200} =$ ________ 4

Question 18: Change the following decimals into fractions.

a 0.21 = ________ **b** 0.85 = ________ **c** 2.25 = ________ **d** 4.25 = ________ 4

Question 19: Change the following fractions into percentages.

a $\frac{9}{100} =$ ________ **b** $\frac{43}{200} =$ ________ **c** $\frac{65}{400} =$ ________ **d** $9\frac{1}{4} =$ ________ 4

Question 20: Change the following percentages into fractions.

a 58% = ________ **b** 69% = ________ **c** 250% = ________ **d** 560% = ________ 4

Question 21:

Complete the following table by filling in the missing fractions, decimals and percentages.

	Fraction	Decimal	Percentage
a	$\frac{23}{100}$		
b		3.25	
c			98%

2

2

2

End of Paper

Total marks for Part C /30

ANSWERS

UNIT 1 **1** P **2** I **3** P **4** I **5** MN **6** P **7** MN **8** P **9** P **10** I **11** MN **12** MN **13** I **14** I **15** P **16** MN **17** P **18** P **19** MN **20** MN **21** MN **22** I **23** P **24** I **25** P **26** MN **27** MN **28** P **29** P **30** I **31** MN **32** I **33** MN **34** P **35** I **36** I **37** MN **38** P **39** I **40** I **41** MN **42** I **43** P **44** MN

UNIT 2 **1** $1\frac{1}{2}$ **2** $2\frac{1}{2}$ **3** $3\frac{1}{2}$ **4** $4\frac{1}{2}$ **5** $1\frac{1}{3}$ **6** $1\frac{2}{3}$ **7** $2\frac{1}{3}$ **8** $3\frac{2}{3}$ **9** $1\frac{3}{4}$ **10** $2\frac{1}{4}$ **11** $2\frac{3}{4}$ **12** $3\frac{1}{4}$ **13** $1\frac{3}{5}$ **14** $1\frac{4}{5}$ **15** $2\frac{1}{5}$ **16** $2\frac{3}{5}$ **17** $12\frac{1}{2}$ **18** $3\frac{4}{5}$ **19** $6\frac{3}{4}$ **20** $6\frac{4}{5}$ **21** $2\frac{1}{6}$ **22** $9\frac{1}{3}$ **23** $2\frac{1}{9}$ **24** $3\frac{2}{5}$ **25** $5\frac{1}{7}$ **26** $4\frac{1}{3}$ **27** $5\frac{3}{5}$ **28** $7\frac{2}{5}$ **29** $7\frac{5}{8}$ **30** $3\frac{8}{9}$ **31** $9\frac{1}{2}$ **32** $10\frac{5}{7}$ **33** $4\frac{4}{7}$ **34** $6\frac{11}{12}$ **35** $7\frac{2}{9}$ **36** $4\frac{4}{11}$ **37** $8\frac{1}{11}$ **38** $15\frac{1}{2}$ **39** $9\frac{5}{9}$ **40** $8\frac{1}{12}$

UNIT 3 **1** $\frac{5}{4}$ **2** $\frac{6}{5}$ **3** $\frac{4}{3}$ **4** $\frac{9}{7}$ **5** $\frac{5}{2}$ **6** $\frac{7}{2}$ **7** $\frac{9}{2}$ **8** $\frac{11}{2}$ **9** $\frac{7}{3}$ **10** $\frac{10}{3}$ **11** $\frac{13}{3}$ **12** $\frac{16}{3}$ **13** $\frac{25}{4}$ **14** $\frac{29}{4}$ **15** $\frac{33}{4}$ **16** $\frac{37}{4}$ **17** $\frac{23}{7}$ **18** $\frac{21}{4}$ **19** $\frac{41}{5}$ **20** $\frac{55}{9}$ **21** $\frac{35}{4}$ **22** $\frac{63}{5}$ **23** $\frac{35}{2}$ **24** $\frac{71}{3}$ **25** $\frac{41}{3}$ **26** $\frac{92}{5}$ **27** $\frac{73}{3}$ **28** $\frac{89}{4}$ **29** $\frac{87}{4}$ **30** $\frac{50}{3}$ **31** $\frac{56}{3}$ **32** $\frac{95}{6}$ **33** $\frac{44}{3}$ **34** $\frac{56}{5}$ **35** $\frac{103}{4}$ **36** $\frac{53}{2}$

UNIT 4 **1** $\frac{1}{2}$ **2** $\frac{3}{4}$ **3** $\frac{4}{5}$ **4** $\frac{6}{7}$ **5** $\frac{1}{3}$ **6** $\frac{2}{5}$ **7** $\frac{3}{7}$ **8** $\frac{5}{7}$ **9** $\frac{1}{4}$ **10** $\frac{1}{6}$ **11** $\frac{2}{7}$ **12** $\frac{3}{8}$ **13** $\frac{4}{9}$ **14** $\frac{5}{12}$ **15** $\frac{6}{11}$ **16** $\frac{7}{13}$ **17** $\frac{8}{11}$ **18** $\frac{10}{13}$ **19** $\frac{14}{15}$ **20** $\frac{21}{25}$ **21** $\frac{1}{2}$ **22** $\frac{1}{3}$ **23** $\frac{1}{4}$ **24** $\frac{1}{5}$ **25** $\frac{2}{3}$ **26** $\frac{4}{5}$ **27** $\frac{6}{7}$ **28** $\frac{8}{9}$ **29** $\frac{1}{8}$ **30** $\frac{2}{5}$ **31** $\frac{3}{7}$ **32** $\frac{4}{9}$ **33** $\frac{5}{6}$ **34** $\frac{6}{11}$ **35** $\frac{7}{8}$ **36** $\frac{8}{13}$ **37** $\frac{9}{10}$ **38** $\frac{11}{12}$ **39** $\frac{13}{15}$ **40** $\frac{15}{17}$

UNIT 5 **1** $\frac{1}{3}$ **2** $\frac{1}{4}$ **3** $\frac{1}{6}$ **4** $\frac{1}{7}$ **5** $\frac{2}{5}$ **6** $\frac{2}{7}$ **7** $\frac{2}{9}$ **8** $\frac{3}{4}$ **9** $\frac{3}{5}$ **10** $\frac{3}{8}$ **11** $\frac{3}{10}$ **12** $\frac{3}{11}$ **13** $\frac{1}{8}$ **14** $\frac{2}{11}$ **15** $\frac{3}{13}$ **16** $\frac{4}{5}$ **17** $\frac{4}{7}$ **18** $\frac{4}{9}$ **19** $\frac{4}{11}$ **20** $\frac{4}{13}$ **21** $\frac{5}{6}$ **22** $\frac{5}{7}$ **23** $\frac{5}{8}$ **24** $\frac{5}{9}$ **25** $\frac{6}{7}$ **26** $\frac{6}{11}$ **27** $\frac{6}{13}$ **28** $\frac{6}{17}$ **29** $\frac{7}{8}$ **30** $\frac{7}{9}$ **31** $\frac{7}{10}$ **32** $\frac{7}{11}$ **33** $\frac{8}{9}$ **34** $\frac{8}{11}$ **35** $\frac{8}{13}$ **36** $\frac{8}{15}$ **37** $\frac{9}{10}$ **38** $\frac{9}{11}$ **39** $\frac{9}{13}$ **40** $\frac{9}{14}$ **41** $\frac{10}{11}$ **42** $\frac{10}{13}$ **43** $\frac{10}{17}$ **44** $\frac{11}{12}$

UNIT 6 **1** $\frac{1}{3}$ **2** $\frac{1}{4}$ **3** $\frac{1}{5}$ **4** $\frac{1}{6}$ **5** $\frac{2}{3}$ **6** $\frac{3}{4}$ **7** $\frac{3}{5}$ **8** $\frac{4}{7}$ **9** $\frac{1}{3}$ **10** $\frac{1}{4}$ **11** $\frac{1}{5}$ **12** $\frac{1}{6}$ **13** $\frac{2}{3}$ **14** $\frac{3}{5}$ **15** $\frac{3}{4}$ **16** $\frac{3}{7}$ **17** $\frac{1}{2}$ **18** $\frac{1}{3}$ **19** $\frac{1}{4}$ **20** $\frac{1}{5}$ **21** $\frac{2}{3}$ **22** $\frac{3}{4}$ **23** $\frac{4}{5}$ **24** $\frac{5}{6}$ **25** $\frac{1}{2}$ **26** $\frac{1}{3}$ **27** $\frac{1}{4}$ **28** $\frac{1}{5}$ **29** $\frac{2}{5}$ **30** $\frac{3}{5}$ **31** $\frac{3}{7}$ **32** $\frac{2}{9}$ **33** $\frac{3}{7}$ **34** $\frac{5}{11}$ **35** $\frac{7}{15}$ **36** $\frac{3}{17}$ **37** $\frac{4}{21}$ **38** $\frac{3}{11}$ **39** $\frac{3}{7}$ **40** $\frac{6}{7}$ **41** $\frac{6}{7}$ **42** $\frac{6}{7}$ **43** $\frac{9}{10}$ **44** $\frac{2}{5}$

UNIT 7 **1** $1\frac{1}{2}$ **2** $1\frac{1}{4}$ **3** $1\frac{1}{3}$ **4** $1\frac{2}{3}$ **5** $2\frac{1}{2}$ **6** $1\frac{1}{5}$ **7** $3\frac{1}{2}$ **8** $2\frac{1}{3}$ **9** $1\frac{3}{4}$ **10** $1\frac{2}{5}$ **11** $1\frac{1}{6}$ **12** $2\frac{2}{3}$ **13** $1\frac{3}{5}$ **14** $1\frac{1}{7}$ **15** $4\frac{1}{2}$ **16** $2\frac{1}{4}$ **17** $\frac{3}{2}$ **18** $\frac{5}{2}$ **19** $\frac{7}{2}$ **20** $\frac{9}{2}$ **21** $\frac{5}{4}$ **22** $\frac{9}{4}$ **23** $\frac{13}{4}$ **24** $\frac{17}{4}$ **25** $\frac{8}{3}$ **26** $\frac{12}{5}$ **27** $\frac{16}{7}$ **28** $\frac{20}{9}$ **29** $\frac{10}{3}$ **30** $\frac{14}{3}$ **31** $\frac{27}{5}$ **32** $\frac{13}{2}$ **33** $\frac{1}{2}$ **34** $\frac{1}{3}$ **35** $\frac{1}{4}$ **36** $\frac{1}{5}$ **37** $\frac{1}{2}$ **38** $\frac{2}{3}$ **39** $\frac{3}{4}$ **40** $\frac{4}{5}$ **41** $\frac{1}{2}$ **42** $\frac{1}{5}$ **43** $\frac{1}{7}$ **44** $\frac{1}{11}$ **45** $\frac{1}{2}$ **46** $\frac{1}{3}$ **47** $\frac{2}{3}$ **48** $\frac{3}{4}$

UNIT 8 **1** P **2** I **3** MN **4** MN **5** P **6** I **7** I **8** I **9** I **10** MN **11** P **12** I **13** $4\frac{3}{4}$ **14** $4\frac{1}{5}$ **15** $5\frac{1}{7}$ **16** $6\frac{2}{9}$ **17** $5\frac{3}{7}$ **18** $6\frac{8}{9}$ **19** $3\frac{1}{8}$ **20** $4\frac{1}{8}$ **21** $9\frac{1}{2}$ **22** $7\frac{7}{8}$ **23** $5\frac{2}{3}$ **24** $5\frac{3}{8}$ **25** $\frac{35}{3}$ **26** $\frac{61}{4}$ **27** $\frac{50}{3}$ **28** $\frac{51}{4}$ **29** $\frac{69}{8}$ **30** $\frac{29}{3}$ **31** $\frac{55}{8}$ **32** $\frac{47}{3}$ **33** $\frac{33}{2}$ **34** $\frac{53}{4}$ **35** $\frac{93}{4}$ **36** $\frac{103}{2}$ **37** $\frac{3}{14}$ **38** $\frac{4}{13}$ **39** $\frac{3}{7}$ **40** $\frac{2}{5}$ **41** $\frac{2}{7}$ **42** $\frac{3}{4}$ **43** $\frac{4}{7}$ **44** $\frac{4}{7}$ **45** $\frac{2}{3}$ **46** $\frac{3}{8}$ **47** $\frac{2}{5}$ **48** $\frac{3}{11}$

UNIT 10

1 **a** I **b** I **c** P **d** I **e** MN **f** I **g** P **h** MN **i** I **j** I **k** P **l** MN **2** **a** $3\frac{3}{5}$ **b** $9\frac{1}{7}$ **c** $3\frac{5}{9}$ **d** $9\frac{1}{11}$ **e** $7\frac{1}{6}$ **f** $7\frac{1}{8}$ **g** $13\frac{2}{3}$ **h** $7\frac{2}{5}$ **i** $11\frac{1}{2}$ **j** $6\frac{7}{9}$ **k** $5\frac{3}{10}$ **l** $5\frac{3}{7}$ **3** **a** $\frac{23}{4}$ **b** $\frac{17}{2}$ **c** $\frac{73}{7}$ **d** $\frac{29}{5}$ **e** $\frac{27}{7}$ **f** $\frac{49}{5}$ **g** $\frac{32}{3}$ **h** $\frac{37}{3}$ **i** $\frac{95}{6}$ **j** $\frac{51}{4}$ **k** $\frac{35}{3}$ **l** $\frac{82}{5}$ **4** **a** $\frac{4}{15}$ **b** $\frac{3}{8}$ **c** $\frac{3}{13}$ **d** $\frac{9}{26}$ **e** $\frac{5}{18}$ **f** $\frac{3}{4}$ **g** $\frac{4}{5}$ **h** $\frac{7}{16}$ **i** $\frac{9}{20}$ **j** $\frac{3}{11}$ **k** $\frac{4}{9}$ **l** $\frac{1}{3}$

UNIT 11

1 6 **2** 6 **3** 20 **4** 15 **5** 9 **6** 25 **7** 12 **8** 30 **9** 8 **10** 15 **11** 28 **12** 16 **13** 14 **14** 18 **15** 25 **16** 49 **17** 27 **18** 35 **19** 63 **20** 72 **21** 40 **22** 45 **23** 27 **24** 48 **25** 22 **26** 36 **27** 56 **28** 75 **29** 28 **30** 27 **31** 44 **32** 35 **33** 45 **34** 42 **35** 91 **36** 33 **37** 52 **38** 75 **39** 57 **40** 45

UNIT 12

1 2 **2** 3 **3** 4 **4** 5 **5** 2 **6** 3 **7** 4 **8** 5 **9** 12 **10** 6 **11** 12 **12** 6 **13** 6 **14** 9 **15** 12 **16** 15 **17** 6 **18** 12 **19** 21 **20** 9 **21** 14 **22** 8 **23** 7 **24** 18 **25** 2 **26** 2 **27** 6 **28** 4 **29** 12 **30** 15 **31** 21 **32** 9 **33** 16 **34** 24 **35** 20 **36** 28 **37** 14 **38** 54 **39** 40 **40** 33

UNIT 13

1 3 **2** 3 **3** 4 **4** 4 **5** 4 **6** 5 **7** 3 **8** 6 **9** 8 **10** 9 **11** 4 **12** 6 **13** 10 **14** 3 **15** 6 **16** 11 **17** 5 **18** 4 **19** 16 **20** 3 **21** 5 **22** 12 **23** 13 **24** 3 **25** 10 **26** 7 **27** 7 **28** 13 **29** 13 **30** 8 **31** 7 **32** 3 **33** 4 **34** 5 **35** 3 **36** 4 **37** 9 **38** 7 **39** 3 **40** 5

UNIT 14

1 5 **2** 4 **3** 3 **4** 5 **5** 1 **6** 1 **7** 2 **8** 1 **9** 1 **10** 4 **11** 1 **12** 4 **13** 5 **14** 8 **15** 2 **16** 4 **17** 8 **18** 7 **19** 8 **20** 4 **21** 9 **22** 2 **23** 1 **24** 5 **25** 3 **26** 5 **27** 7 **28** 3 **29** 3 **30** 6 **31** 2 **32** 3 **33** 2 **34** 2 **35** 4 **36** 9 **37** 3 **38** 2 **39** 2 **40** 7 **41** 2 **42** 7 **43** 17 **44** 19

UNIT 15

1 $<$ **2** $>$ **3** $>$ **4** $<$ **5** $>$ **6** $>$ **7** $<$ **8** $>$ **9** $<$ **10** $>$ **11** $>$ **12** $<$ **13** $<$ **14** $>$ **15** $<$ **16** $>$ **17** $>$ **18** $>$ **19** $<$ **20** $>$ **21** $\frac{1}{4}, \frac{2}{4}, \frac{3}{4}$ **22** $\frac{1}{5}, \frac{3}{5}, \frac{4}{5}$ **23** $\frac{1}{7}, \frac{2}{7}, \frac{3}{7}$ **24** $\frac{2}{9}, \frac{6}{9}, \frac{8}{9}$ **25** $\frac{2}{11}, \frac{3}{11}, \frac{7}{11}$ **26** $\frac{2}{13}, \frac{5}{13}, \frac{7}{13}$ **27** $\frac{2}{21}, \frac{3}{21}, \frac{8}{21}$ **28** $\frac{3}{25}, \frac{6}{25}, \frac{11}{25}$ **29** $\frac{1}{17}, \frac{2}{17}, \frac{8}{17}$ **30** $\frac{2}{15}, \frac{5}{15}, \frac{6}{15}$ **31** $\frac{1}{14}, \frac{3}{14}, \frac{8}{14}$ **32** $\frac{1}{13}, \frac{3}{13}, \frac{9}{13}$ **33** $\frac{1}{26}, \frac{5}{26}, \frac{9}{26}$ **34** $\frac{1}{29}, \frac{2}{29}, \frac{3}{29}$ **35** $\frac{2}{31}, \frac{5}{31}, \frac{7}{31}$

UNIT 16

1 $<$ **2** $>$ **3** $>$ **4** $<$ **5** $<$ **6** $<$ **7** $<$ **8** $>$ **9** $<$ **10** $<$ **11** $<$ **12** $<$ **13** $<$ **14** $>$ **15** $>$ **16** $>$ **17** $>$ **18** $<$

UNIT 17

1 $<$ **2** $>$ **3** $>$ **4** $>$ **5** $<$ **6** $>$ **7** $<$ **8** $<$ **9** $<$ **10** $<$ **11** $<$ **12** $<$ **13** $<$ **14** $<$ **15** $>$

UNIT 18

1 $>$ **2** $>$ **3** $<$ **4** $<$ **5** $>$ **6** $=$ **7** $>$ **8** $=$ **9** $>$ **10** $>$ **11** $=$ **12** $=$ **13** $>$ **14** $>$ **15** $>$ **16** $>$ **17** $<$ **18** $>$ **19** $<$ **20** $>$ **21** $\frac{1}{3}, \frac{1}{2}, \frac{2}{3}$ **22** $\frac{4}{5}, \frac{5}{6}, \frac{6}{7}$ **23** $\frac{3}{5}, \frac{3}{4}, \frac{3}{3}$ **24** $\frac{8}{20}, \frac{7}{10}, \frac{80}{100}$ **25** $\frac{1}{8}, \frac{1}{5}, \frac{1}{3}$ **26** $\frac{3}{8}, \frac{2}{4}, \frac{3}{4}$ **27** $\frac{3}{10}, \frac{2}{4}, \frac{3}{5}$ **28** $\frac{1}{9}, \frac{1}{6}, \frac{1}{3}$ **29** $\frac{7}{20}, \frac{4}{5}, \frac{9}{10}$ **30** $\frac{3}{8}, \frac{2}{5}, \frac{4}{7}$ **31** $\frac{1}{4}, \frac{2}{5}, \frac{3}{7}$ **32** $\frac{4}{9}, \frac{3}{5}, \frac{3}{4}$

UNIT 19

1 4 **2** 6 **3** 8 **4** 10 **5** 9 **6** 15 **7** 21 **8** 27 **9** 3 **10** 2 **11** 2 **12** 4 **13** 4 **14** 9 **15** 12 **16** 20 **17** 3 **18** 4 **19** 5 **20** 6 **21** 2 **22** 3 **23** 3 **24** 3 **25** 2 **26** 1 **27** 1 **28** 1 **29** 2 **30** 3 **31** 3 **32** 5 **33** $\frac{1}{6}, \frac{3}{6}, \frac{6}{6}$ **34** $\frac{1}{5}, \frac{2}{5}, \frac{3}{5}$ **35** $\frac{1}{7}, \frac{4}{7}, \frac{6}{7}$ **36** $\frac{1}{9}, \frac{2}{9}, \frac{5}{9}$ **37** $\frac{5}{11}, \frac{6}{11}, \frac{8}{11}$ **38** $\frac{1}{3}, \frac{2}{3}, \frac{3}{3}$ **39** $\frac{3}{12}, \frac{7}{12}, \frac{8}{12}$ **40** $\frac{2}{13}, \frac{4}{13}, \frac{5}{13}$ **41** $\frac{3}{9}, \frac{5}{9}, \frac{6}{9}$ **42** $\frac{2}{14}, \frac{8}{14}, \frac{9}{14}$ **43** $\frac{1}{15}, \frac{2}{15}, \frac{3}{15}$ **44** $\frac{5}{16}, \frac{7}{16}, \frac{8}{16}$ **45** $\frac{1}{4}, \frac{3}{4}, \frac{4}{4}$ **46** $\frac{3}{8}, \frac{4}{8}, \frac{5}{8}$ **47** $\frac{7}{10}, \frac{9}{10}, \frac{10}{10}$

UNIT 20

1 21 **2** 20 **3** 36 **4** 55 **5** 72 **6** 105 **7** 84 **8** 48 **9** 16 **10** 27 **11** 24 **12** 30 **13** 54 **14** 63 **15** 72 **16** 60 **17** 9 **18** 12 **19** 3 **20** 20 **21** 25 **22** 10 **23** 7 **24** 41 **25** 2 **26** 11 **27** 5 **28** 15 **29** 10 **30** 25 **31** 8 **32** 15

33 $\frac{5}{7}, \frac{2}{7}, \frac{1}{7}$ **34** $\frac{3}{4}, \frac{8}{12}, \frac{7}{12}$ **35** $\frac{9}{21}, \frac{7}{21}, \frac{3}{21}$ **36** $\frac{10}{15}, \frac{8}{15}, \frac{2}{5}$ **37** $\frac{2}{3}, \frac{3}{8}, \frac{1}{4}$ **38** $\frac{3}{4}, \frac{2}{3}, \frac{6}{12}$ **39** $\frac{3}{4}, \frac{5}{8}, \frac{1}{2}$ **40** $\frac{8}{5}, \frac{15}{20}, \frac{6}{10}$
41 $\frac{5}{8}, \frac{8}{24}, \frac{1}{4}$ **42** $\frac{17}{30}, \frac{2}{5}, \frac{1}{6}$ **43** $\frac{4}{5}, \frac{8}{15}, \frac{3}{10}$ **44** $\frac{8}{10}, \frac{2}{5}, \frac{3}{15}$ **45** $\frac{8}{9}, \frac{7}{18}, \frac{1}{3}$ **46** $\frac{8}{21}, \frac{5}{14}, \frac{2}{7}$ **47** $\frac{7}{8}, \frac{3}{4}, \frac{3}{8}$

UNIT 22 **1 a** 21 **b** 36 **c** 42 **d** 75 **e** 36 **f** 72 **g** 56 **h** 80 **2 a** 8 **b** 18 **c** 40 **d** 63 **e** 27 **f** 50 **g** 45 **h** 48 **3 a** 3 **b** 4 **c** 4 **d** 5 **e** 3 **f** 3 **g** 3 **h** 8
4 a $\frac{8}{11}, \frac{5}{11}, \frac{2}{11}$ **b** $\frac{5}{21}, \frac{3}{21}, \frac{2}{21}$ **c** $\frac{8}{37}, \frac{5}{37}, \frac{3}{37}$ **d** $\frac{5}{6}, \frac{2}{3}, \frac{4}{12}$ **e** $\frac{7}{15}, \frac{2}{5}, \frac{1}{15}$ **f** $\frac{3}{4}, \frac{7}{16}, \frac{3}{8}$
5 a $\frac{1}{8}, \frac{2}{8}, \frac{3}{8}$ **b** $\frac{3}{5}, \frac{5}{11}, \frac{9}{11}$ **c** $\frac{3}{15}, \frac{6}{15}, \frac{7}{15}$ **d** $\frac{1}{2}, \frac{3}{4}, \frac{7}{8}$ **e** $\frac{7}{12}, \frac{2}{3}, \frac{5}{6}$ **f** $\frac{1}{3}, \frac{5}{9}, \frac{5}{6}$ **g** $\frac{5}{8}, \frac{3}{4}, \frac{11}{22}$ **h** $\frac{3}{10}, \frac{1}{2}, \frac{3}{5}$ **i** $\frac{3}{8}, \frac{3}{4}, \frac{5}{6}$

UNIT 23 **1** $\frac{2}{3}$ **2** $\frac{3}{5}$ **3** $\frac{6}{7}$ **4** $\frac{7}{9}$ **5** $\frac{10}{11}$ **6** $\frac{7}{10}$ **7** $\frac{3}{4}$ **8** $\frac{5}{8}$ **9** $\frac{11}{15}$ **10** $\frac{7}{12}$ **11** $\frac{9}{17}$ **12** $\frac{11}{21}$ **13** $\frac{1}{3}$ **14** $\frac{1}{2}$ **15** $\frac{2}{3}$ **16** $\frac{3}{5}$ **17** $\frac{1}{2}$ **18** $\frac{3}{5}$ **19** $\frac{1}{3}$ **20** $\frac{3}{5}$ **21** $\frac{4}{5}$ **22** $\frac{3}{4}$ **23** $\frac{3}{4}$ **24** $\frac{2}{3}$

UNIT 24 **1** $\frac{9}{10}$ **2** $\frac{7}{8}$ **3** $1\frac{3}{10}$ **4** $\frac{5}{6}$ **5** $\frac{7}{12}$ **6** $\frac{5}{14}$ **7** $\frac{11}{20}$ **8** $\frac{11}{15}$ **9** $\frac{8}{9}$ **10** $\frac{13}{22}$ **11** $\frac{17}{24}$ **12** $\frac{11}{18}$ **13** $\frac{2}{3}$ **14** $\frac{3}{4}$ **15** $\frac{5}{6}$ **16** $\frac{7}{8}$ **17** $\frac{8}{9}$ **18** $\frac{7}{12}$ **19** $1\frac{1}{3}$ **20** $\frac{2}{3}$ **21** $\frac{3}{4}$

UNIT 25 **1** $1\frac{2}{3}$ **2** $\frac{11}{20}$ **3** $1\frac{1}{7}$ **4** $1\frac{1}{20}$ **5** $1\frac{4}{15}$ **6** $\frac{14}{15}$ **7** $1\frac{5}{8}$ **8** $2\frac{5}{24}$ **9** $1\frac{47}{90}$ **10** $1\frac{11}{12}$ **11** $2\frac{1}{20}$ **12** $1\frac{23}{24}$

UNIT 26 **1** $10\frac{3}{4}$ **2** 10 **3** $1\frac{5}{24}$ **4** $8\frac{11}{12}$ **5** $2\frac{1}{8}$ **6** $10\frac{1}{4}$ **7** $8\frac{2}{5}$ **8** 4 **9** $8\frac{3}{4}$ **10** 7 **11** $4\frac{5}{6}$ **12** $5\frac{3}{4}$ **13** $4\frac{1}{20}$ **14** $7\frac{3}{20}$ **15** $8\frac{1}{3}$ **16** $3\frac{4}{9}$ **17** $8\frac{3}{4}$ **18** $11\frac{1}{12}$

UNIT 27 **1** $\frac{1}{3}$ **2** $\frac{2}{7}$ **3** $\frac{2}{9}$ **4** $\frac{2}{13}$ **5** $\frac{3}{17}$ **6** $\frac{3}{25}$ **7** $\frac{8}{35}$ **8** $\frac{2}{15}$ **9** $\frac{5}{42}$ **10** $\frac{5}{37}$ **11** $\frac{1}{3}$ **12** $\frac{1}{2}$ **13** $\frac{1}{3}$ **14** $\frac{1}{9}$ **15** $\frac{1}{3}$ **16** $\frac{1}{3}$ **17** $\frac{1}{5}$ **18** $\frac{1}{4}$ **19** $\frac{2}{5}$ **20** $\frac{1}{7}$ **21** $\frac{1}{4}$ **22** $\frac{1}{5}$ **23** $\frac{3}{8}$ **24** $\frac{1}{9}$

UNIT 28 **1** $\frac{3}{10}$ **2** $\frac{2}{5}$ **3** $\frac{1}{20}$ **4** $\frac{7}{10}$ **5** $\frac{1}{5}$ **6** $\frac{5}{24}$ **7** $\frac{9}{50}$ **8** $\frac{2}{5}$ **9** $\frac{7}{15}$ **10** $\frac{1}{4}$ **11** $\frac{2}{5}$ **12** $\frac{1}{6}$ **13** $\frac{1}{2}$ **14** $\frac{1}{7}$ **15** $\frac{19}{20}$ **16** $\frac{4}{15}$ **17** $\frac{7}{12}$ **18** $\frac{20}{63}$ **19** $\frac{11}{21}$ **20** $\frac{5}{14}$ **21** $\frac{23}{25}$

UNIT 29 **1** $2\frac{5}{12}$ **2** $4\frac{2}{5}$ **3** $\frac{1}{2}$ **4** $7\frac{1}{30}$ **5** $1\frac{1}{6}$ **6** $1\frac{7}{12}$ **7** $5\frac{1}{12}$ **8** $5\frac{7}{9}$ **9** $2\frac{5}{9}$ **10** $3\frac{1}{4}$ **11** $3\frac{13}{20}$ **12** $1\frac{17}{36}$ **13** $3\frac{1}{6}$ **14** $6\frac{1}{12}$ **15** $\frac{13}{24}$

UNIT 30 **1** 1 **2** $\frac{2}{3}$ **3** $\frac{2}{5}$ **4** $\frac{2}{3}$ **5** $\frac{3}{4}$ **6** $1\frac{1}{6}$ **7** 1 **8** $\frac{9}{20}$ **9** $\frac{5}{6}$ **10** 1 **11** $\frac{1}{3}$ **12** $\frac{1}{6}$ **13** $\frac{7}{9}$ **14** $\frac{1}{4}$ **15** $\frac{4}{5}$ **16** $\frac{8}{15}$ **17** $\frac{7}{12}$ **18** $\frac{3}{5}$ **19** $1\frac{1}{5}$ **20** $\frac{13}{20}$ **21** $1\frac{1}{3}$ **22** $\frac{13}{14}$ **23** $\frac{2}{9}$ **24** $\frac{5}{12}$ **25** $\frac{1}{8}$ **26** $\frac{1}{2}$

UNIT 31 **1** 1 **2** $\frac{73}{84}$ **3** $\frac{73}{88}$ **4** $1\frac{1}{3}$ **5** $1\frac{1}{18}$ **6** $\frac{17}{33}$ **7** $\frac{13}{24}$ **8** $\frac{7}{45}$ **9** $\frac{4}{9}$ **10** $\frac{3}{8}$ **11** $\frac{4}{55}$ **12** $\frac{25}{66}$ **13** $\frac{16}{33}$ **14** $\frac{7}{36}$ **15** $6\frac{13}{21}$ **16** $11\frac{67}{88}$ **17** $8\frac{19}{30}$ **18** $8\frac{61}{66}$ **19** $10\frac{2}{3}$ **20** $2\frac{13}{20}$ **21** $2\frac{1}{2}$ **22** $4\frac{17}{40}$ **23** $3\frac{3}{5}$ **24** $4\frac{2}{3}$ **25** $9\frac{8}{9}$ **26** $5\frac{5}{6}$

UNIT 33 **1 a** $\frac{3}{7}$ **b** $\frac{2}{3}$ **c** $\frac{25}{72}$ **d** $\frac{23}{30}$ **e** $1\frac{12}{55}$ **f** $\frac{115}{132}$ **g** $1\frac{7}{18}$ **h** $1\frac{11}{12}$ **i** $13\frac{1}{4}$ **j** $11\frac{3}{4}$ **k** $6\frac{1}{12}$ **l** $8\frac{5}{8}$ **2 a** $\frac{5}{11}$ **b** $\frac{3}{7}$ **c** $\frac{5}{18}$ **d** $\frac{1}{9}$ **e** $\frac{53}{88}$ **f** $\frac{34}{65}$ **g** $2\frac{1}{4}$ **h** $1\frac{4}{5}$ **i** $1\frac{43}{70}$ **j** $3\frac{5}{14}$ **k** $10\frac{2}{15}$ **l** $7\frac{19}{20}$ **m** 2 **n** $5\frac{5}{28}$

UNIT 34 **1** $\frac{1}{4}$ **2** $\frac{1}{12}$ **3** $\frac{1}{20}$ **4** $\frac{1}{30}$ **5** $\frac{2}{15}$ **6** $\frac{3}{20}$ **7** $\frac{2}{21}$ **8** $\frac{6}{35}$ **9** $\frac{1}{30}$ **10** $\frac{6}{55}$ **11** $\frac{25}{48}$ **12** $\frac{10}{63}$ **13** $\frac{18}{35}$ **14** $\frac{15}{52}$ **15** $\frac{24}{35}$ **16** $\frac{35}{88}$ **17** $\frac{27}{34}$ **18** $\frac{27}{50}$ **19** $\frac{16}{81}$ **20** $\frac{6}{55}$ **21** $\frac{5}{18}$ **22** $\frac{24}{91}$ **23** $\frac{12}{49}$ **24** $\frac{32}{81}$ **25** $\frac{10}{27}$ **26** $\frac{35}{72}$ **27** $\frac{27}{32}$ **28** $\frac{14}{45}$ **29** $\frac{46}{75}$ **30** $\frac{14}{39}$ **31** $\frac{56}{81}$ **32** $\frac{48}{77}$ **33** $\frac{25}{33}$ **34** $\frac{7}{32}$ **35** $\frac{18}{35}$ **36** $\frac{21}{40}$ **37** $\frac{21}{100}$ **38** $\frac{3}{100}$ **39** $\frac{48}{55}$ **40** $\frac{35}{96}$ **41** $\frac{48}{85}$ **42** $\frac{65}{84}$ **43** $1\frac{11}{49}$ **44** $\frac{16}{99}$ **45** $\frac{50}{99}$ **46** $\frac{48}{65}$ **47** $\frac{32}{99}$ **48** $\frac{63}{77}$

UNIT 35 **1** $\frac{1}{10}$ **2** $\frac{3}{7}$ **3** $\frac{7}{18}$ **4** $\frac{5}{12}$ **5** $\frac{10}{27}$ **6** $\frac{6}{13}$ **7** $\frac{1}{4}$ **8** $\frac{4}{13}$ **9** 1 **10** $\frac{3}{4}$ **11** $\frac{1}{10}$ **12** $\frac{7}{20}$ **13** $\frac{2}{3}$ **14** $\frac{1}{2}$ **15** $\frac{2}{3}$ **16** $\frac{2}{5}$ **17** $\frac{1}{3}$ **18** $\frac{3}{16}$

UNIT 36 **1** $\frac{1}{3}$ **2** $\frac{5}{12}$ **3** $\frac{1}{10}$ **4** $\frac{3}{7}$ **5** $\frac{1}{7}$ **6** $\frac{3}{13}$ **7** $\frac{6}{13}$ **8** $\frac{5}{7}$ **9** $\frac{6}{17}$ **10** $\frac{24}{35}$ **11** $\frac{4}{45}$ **12** $\frac{6}{25}$ **13** $\frac{6}{23}$ **14** $\frac{9}{25}$ **15** $\frac{20}{29}$ **16** $\frac{4}{5}$ **17** $\frac{2}{3}$ **18** $\frac{1}{2}$ **19** $\frac{2}{3}$ **20** 2 **21** $\frac{4}{5}$ **22** $\frac{1}{7}$ **23** $\frac{8}{15}$ **24** $\frac{1}{14}$ **25** 4 **26** $\frac{1}{2}$ **27** $\frac{1}{2}$ **28** $\frac{4}{7}$

UNIT 37 **1** $3\frac{3}{8}$ **2** 8 **3** $6\frac{1}{2}$ **4** $2\frac{3}{16}$ **5** $5\frac{5}{8}$ **6** $18\frac{3}{8}$ **7** $7\frac{1}{2}$ **8** $3\frac{3}{20}$ **9** $4\frac{7}{8}$ **10** $15\frac{3}{16}$ **11** $11\frac{13}{25}$ **12** $5\frac{2}{15}$ **13** $9\frac{11}{12}$ **14** $9\frac{3}{8}$ **15** $7\frac{13}{50}$ **16** $10\frac{1}{5}$ **17** $12\frac{1}{10}$ **18** $13\frac{21}{32}$ **19** $17\frac{17}{28}$ **20** $6\frac{8}{15}$ **21** $56\frac{4}{7}$ **22** $5\frac{5}{6}$ **23** 22 **24** $8\frac{1}{8}$ **25** $28\frac{7}{8}$ **26** $6\frac{1}{4}$ **27** $5\frac{15}{16}$ **28** $47\frac{1}{4}$

UNIT 38 **1** 4 **2** $\frac{4}{9}$ **3** $\frac{7}{10}$ **4** $\frac{4}{15}$ **5** $\frac{5}{8}$ **6** $2\frac{1}{2}$ **7** $2\frac{1}{3}$ **8** $1\frac{4}{5}$ **9** $\frac{2}{3}$ **10** $\frac{2}{7}$ **11** $\frac{2}{9}$ **12** $\frac{4}{21}$ **13** $1\frac{2}{3}$ **14** 5 **15** $\frac{5}{21}$ **16** $\frac{4}{25}$ **17** $1\frac{3}{5}$ **18** $3\frac{2}{3}$ **19** 2 **20** $\frac{2}{9}$ **21** $\frac{5}{13}$ **22** $\frac{1}{9}$ **23** $\frac{7}{20}$ **24** $\frac{2}{19}$ **25** $1\frac{1}{5}$ **26** $3\frac{1}{2}$ **27** $1\frac{1}{9}$ **28** $\frac{3}{10}$ **29** $\frac{4}{9}$ **30** $\frac{2}{11}$ **31** $\frac{4}{5}$ **32** $6\frac{1}{2}$ **33** $\frac{4}{23}$ **34** $\frac{7}{34}$ **35** $\frac{6}{23}$ **36** $\frac{4}{35}$ **37** $\frac{5}{6}$ **38** $\frac{7}{26}$ **39** $\frac{7}{38}$ **40** $\frac{4}{39}$

UNIT 39 **1** $\frac{3}{5}$ **2** $3\frac{1}{2}$ **3** $\frac{2}{3}$ **4** $\frac{8}{15}$ **5** $\frac{15}{16}$ **6** $\frac{25}{36}$ **7** $\frac{6}{25}$ **8** $1\frac{7}{8}$ **9** $2\frac{2}{5}$ **10** $1\frac{1}{32}$ **11** $\frac{8}{9}$ **12** $\frac{32}{35}$ **13** $\frac{99}{100}$ **14** $3\frac{1}{7}$ **15** $\frac{45}{49}$ **16** $\frac{35}{48}$ **17** $\frac{48}{49}$ **18** $2\frac{7}{24}$ **19** $1\frac{1}{10}$ **20** $\frac{80}{121}$ **21** $\frac{18}{35}$ **22** $1\frac{1}{5}$ **23** $4\frac{3}{8}$ **24** $1\frac{19}{35}$

UNIT 40 **1** $\frac{7}{20}$ **2** $1\frac{5}{9}$ **3** $1\frac{1}{2}$ **4** 2 **5** $1\frac{1}{2}$ **6** 1 **7** $\frac{1}{6}$ **8** 6 **9** $\frac{7}{8}$ **10** $\frac{10}{21}$ **11** 1 **12** $\frac{3}{4}$ **13** $\frac{3}{4}$ **14** $\frac{1}{3}$ **15** $\frac{4}{11}$ **16** 1 **17** 1 **18** $\frac{3}{5}$

UNIT 41 **1** $1\frac{5}{9}$ **2** $1\frac{1}{4}$ **3** $\frac{2}{3}$ **4** $\frac{3}{4}$ **5** $1\frac{1}{2}$ **6** $2\frac{1}{4}$ **7** $1\frac{1}{2}$ **8** $\frac{1}{7}$ **9** 6 **10** $\frac{20}{21}$ **11** $4\frac{1}{2}$ **12** 1 **13** $\frac{3}{4}$ **14** 4 **15** $\frac{14}{15}$ **16** $\frac{28}{121}$ **17** $\frac{1}{3}$ **18** 2 **19** $\frac{7}{12}$ **20** $2\frac{26}{243}$ **21** $\frac{96}{121}$ **22** $\frac{1}{2}$ **23** $\frac{9}{10}$ **24** 2

UNIT 42 **1** $1\frac{1}{5}$ **2** $1\frac{13}{17}$ **3** $\frac{16}{169}$ **4** $3\frac{8}{19}$ **5** $\frac{19}{24}$ **6** $8\frac{2}{5}$ **7** $3\frac{171}{175}$ **8** $1\frac{2}{5}$ **9** $\frac{2}{3}$ **10** $1\frac{3}{14}$ **11** $\frac{21}{52}$ **12** $\frac{35}{41}$ **13** $3\frac{6}{7}$ **14** $\frac{3}{4}$ **15** $4\frac{7}{12}$ **16** $2\frac{5}{36}$ **17** $1\frac{4}{17}$ **18** $\frac{27}{40}$

UNIT 43 **1** $\frac{1}{25}$ **2** $\frac{1}{21}$ **3** $\frac{1}{30}$ **4** $\frac{2}{7}$ **5** $\frac{10}{13}$ **6** $\frac{32}{39}$ **7** $\frac{25}{49}$ **8** $\frac{35}{54}$ **9** $\frac{16}{21}$ **10** $\frac{5}{12}$ **11** $\frac{1}{5}$ **12** $\frac{1}{4}$ **13** $\frac{6}{7}$ **14** $\frac{1}{9}$ **15** $\frac{4}{15}$ **16** $3\frac{9}{10}$ **17** $2\frac{13}{56}$ **18** $10\frac{5}{16}$ **19** $1\frac{1}{3}$ **20** $1\frac{1}{5}$ **21** $1\frac{3}{8}$ **22** $\frac{1}{5}$ **23** $1\frac{1}{2}$ **24** 1 **25** 2 **26** $2\frac{1}{12}$ **27** $\frac{1}{2}$

UNIT 44 **1** $\frac{1}{4}$ **2** $\frac{1}{9}$ **3** $\frac{1}{16}$ **4** $\frac{2}{15}$ **5** $\frac{3}{20}$ **6** $\frac{5}{18}$ **7** $\frac{4}{15}$ **8** $\frac{6}{35}$ **9** $\frac{12}{35}$ **10** $\frac{7}{12}$ **11** $\frac{2}{3}$ **12** $\frac{2}{3}$ **13** $\frac{4}{9}$ **14** $\frac{3}{8}$ **15** $\frac{2}{3}$ **16** 40 **17** 15 **18** 40 **19** $4\frac{3}{8}$ **20** 8 **21** $4\frac{1}{10}$ **22** 1 **23** 1 **24** 1 **25** $1\frac{1}{2}$ **26** $1\frac{2}{3}$ **27** $1\frac{1}{2}$ **28** $4\frac{1}{2}$ **29** $1\frac{1}{3}$ **30** $\frac{2}{3}$ **31** 4 **32** 2 **33** 1 **34** $\frac{1}{4}$ **35** $\frac{25}{32}$ **36** $1\frac{17}{81}$ **37** $\frac{5}{8}$ **38** 6 **39** $\frac{5}{8}$ **40** $\frac{1}{2}$ **41** $\frac{1}{2}$ **42** 14 **43** 1 **44** 1 **45** $\frac{11}{16}$

UNIT 45 **1** $4\frac{1}{2}$ **2** $4\frac{1}{3}$ **3** $5\frac{1}{2}$ **4** $6\frac{1}{4}$ **5** $5\frac{5}{6}$ **6** $7\frac{4}{5}$ **7** $7\frac{1}{5}$ **8** $6\frac{6}{7}$ **9** $\frac{13}{4}$ **10** $\frac{35}{6}$ **11** $\frac{29}{3}$ **12** $\frac{71}{6}$ **13** $\frac{26}{3}$ **14** $\frac{68}{7}$
15 $\frac{34}{3}$ **16** $\frac{55}{7}$ **17** $\frac{2}{9}$ **18** $\frac{1}{8}$ **19** $\frac{6}{25}$ **20** $\frac{15}{29}$ **21** 40 **22** 25 **23** 33 **24** 4 **25** 4 **26** 4 **27** 4 **28** 3 **29** $\frac{1}{5}, \frac{2}{5}, \frac{4}{5}$
30 $\frac{1}{4}, \frac{1}{2}, \frac{2}{3}$ **31** $\frac{1}{4}, \frac{1}{3}, \frac{2}{3}$ **32** $\frac{1}{2}, \frac{3}{4}, \frac{5}{6}$ **33** $\frac{1}{4}, \frac{8}{12}, \frac{7}{8}$ **34** $\frac{1}{3}, \frac{2}{3}, \frac{5}{6}$ **35** $\frac{5}{9}$ **36** $\frac{3}{4}$ **37** $\frac{1}{6}$ **38** $5\frac{3}{4}$ **39** $5\frac{17}{30}$ **40** $2\frac{3}{4}$

UNIT 47 **1 a** $\frac{1}{36}$ **b** $\frac{12}{35}$ **c** $\frac{2}{3}$ **d** $\frac{15}{56}$ **e** $\frac{12}{35}$ **f** $\frac{8}{15}$ **g** $\frac{1}{8}$ **h** $\frac{1}{2}$ **i** $3\frac{1}{9}$ **j** $\frac{27}{200}$ **k** $\frac{1}{2}$ **l** $3\frac{5}{6}$
2 a $1\frac{1}{4}$ **b** 1 **c** $\frac{2}{3}$ **d** $\frac{1}{4}$ **e** $\frac{7}{10}$ **f** $\frac{25}{98}$ **g** 1 **h** $11\frac{3}{7}$ **i** $2\frac{2}{27}$ **j** $1\frac{11}{34}$ **k** $\frac{2}{3}$ **l** $\frac{4}{11}$ **m** $2\frac{5}{6}$ **n** 79

UNIT 48 **1** 1 **2** 3 **3** 4 **4** 5 **5** 7 **6** 8 **7** 7 **8** 4 **9** 2 **10** 6 **11** 6 **12** 4 **13** 1 **14** 61 **15** 7 **16** 6
17 6 **18** 2 **19** 6 **20** 10 **21** 9 **22** 62 **23** 16 **24** 13 **25** 6, $5\frac{3}{4}$ **26** 7, 7 **27** 9, $8\frac{1}{4}$ **28** 11, $10\frac{1}{3}$ **29** 6, $5\frac{7}{8}$
30 13, $12\frac{1}{5}$ **31** 6, $6\frac{2}{9}$ **32** 4, $4\frac{13}{36}$ **33** 5, $5\frac{4}{21}$ **34** 3, $2\frac{11}{12}$ **35** 4, $3\frac{1}{3}$ **36** 3, $2\frac{83}{100}$ **37** 4, $5\frac{1}{4}$ **38** 8, $7\frac{3}{4}$ **39** 3, $3\frac{1}{6}$

UNIT 49 **1** 2 **2** 2 **3** 6 **4** 12 **5** 1 **6** 2 **7** 1 **8** 2 **9** 20 **10** 12 **11** 6 **12** 4 **13** 50 **14** 16 **15** 40
16 6 **17** 90 **18** 22 **19** 36 **20** 18 **21** 200 **22** 18 **23** 15 **24** 400 **25** 24 **26** 11 **27** 5 **28** 80 **29** 20
30 16 **31** 400 **32** 400 **33** 100 **34** 350 **35** 200 **36** 80 **37** 200 **38** 500 **39** 550 **40** 90 **41** 30 **42** 80
43 20 **44** 9 **45** 6 **46** 500 **47** 55 **48** 60

UNIT 50 **1** $\frac{1}{25}$ **2** $\frac{1}{9}$ **3** $\frac{1}{16}$ **4** $\frac{1}{36}$ **5** $\frac{4}{25}$ **6** $\frac{9}{25}$ **7** $\frac{16}{49}$ **8** $\frac{9}{64}$ **9** $\frac{16}{81}$ **10** $\frac{25}{49}$ **11** $\frac{36}{121}$ **12** $\frac{49}{81}$ **13** $\frac{4}{49}$ **14** $\frac{9}{64}$ **15** $\frac{25}{81}$
16 $\frac{36}{49}$ **17** $2\frac{1}{4}$ **18** $6\frac{1}{4}$ **19** $5\frac{1}{16}$ **20** $12\frac{1}{4}$ **21** $\frac{1}{4}$ **22** $\frac{1}{9}$ **23** $\frac{1}{16}$ **24** $\frac{1}{25}$ **25** $\frac{4}{9}$ **26** $\frac{4}{25}$ **27** $\frac{4}{49}$ **28** $\frac{4}{81}$ **29** $\frac{9}{25}$ **30** $\frac{9}{49}$
31 $\frac{16}{49}$ **32** $\frac{25}{49}$ **33** $\frac{1}{64}$ **34** $\frac{9}{64}$ **35** $\frac{25}{64}$ **36** $\frac{49}{64}$ **37** $\frac{1}{81}$ **38** $\frac{4}{81}$ **39** $\frac{9}{100}$ **40** $\frac{81}{100}$ **41** $2\frac{1}{4}$ **42** $6\frac{1}{4}$ **43** $12\frac{1}{4}$ **44** $28\frac{4}{9}$

UNIT 51 **1** $\frac{1}{2}$ **2** $\frac{1}{3}$ **3** $\frac{1}{4}$ **4** $\frac{1}{5}$ **5** $\frac{1}{6}$ **6** $\frac{1}{7}$ **7** $\frac{1}{8}$ **8** $\frac{1}{9}$ **9** $\frac{2}{3}$ **10** $\frac{3}{4}$ **11** $\frac{4}{5}$ **12** $\frac{2}{3}$ **13** $\frac{3}{8}$ **14** $\frac{5}{6}$ **15** $\frac{6}{7}$ **16** $\frac{5}{9}$ **17** $\frac{2}{5}$
18 $\frac{5}{11}$ **19** $\frac{1}{2}$ **20** $\frac{2}{9}$ **21** $\frac{3}{11}$ **22** $\frac{5}{6}$ **23** $\frac{11}{25}$ **24** $\frac{4}{5}$ **25** $\frac{2}{3}$ **26** $\frac{4}{9}$ **27** $\frac{3}{5}$ **28** $\frac{6}{7}$ **29** $\frac{8}{9}$ **30** $\frac{4}{7}$ **31** $\frac{1}{2}$ **32** $\frac{11}{15}$ **33** $\frac{1}{3}$
34 $\frac{1}{2}$ **35** $\frac{1}{2}$ **36** $\frac{1}{5}$ **37** $\frac{2}{5}$ **38** $\frac{3}{8}$ **39** $\frac{1}{3}$ **40** $\frac{2}{5}$ **41** $\frac{5}{9}$ **42** $\frac{5}{11}$ **43** $\frac{5}{12}$ **44** $\frac{6}{11}$ **45** $\frac{6}{13}$ **46** $\frac{7}{11}$ **47** $\frac{2}{3}$ **48** $\frac{9}{13}$

UNIT 52 **1** $\frac{11}{12}$ **2** $\frac{23}{30}$ **3** $4\frac{5}{6}$ **4** $\frac{31}{42}$ **5** $7\frac{1}{2}$ **6** $2\frac{1}{15}$ **7** $2\frac{5}{12}$ **8** $5\frac{3}{4}$ **9** $27\frac{1}{8}$ **10** $1\frac{5}{8}$ **11** $7\frac{37}{56}$ **12** $2\frac{9}{10}$ **13** $2\frac{31}{80}$
14 $3\frac{7}{11}$

UNIT 53 **1** $\frac{2}{3}$ **2** $\frac{4}{5}$ **3** 21 **4** $\frac{13}{24}$ **5** $8\frac{1}{4}$ **6** $3\frac{11}{21}$ **7** $5\frac{3}{7}$ **8** $1\frac{7}{24}$ **9** $5\frac{4}{9}$

UNIT 54 **1** \$4 **2** \$10 **3** \$8 **4** \$22 **5** \$75 **6** 50 min **7** 4 min **8** 900 g **9** 90 years **10** 30 cm
11 7 weeks **12** 600 mL **13** 100 m **14** 50 g **15** 50 mL **16** 1500 L **17** \$2.20 **18** 15 min **19** 2700 g
20 160 cm **21** \$81 **22** \$30 **23** 16 hours **24** 27 km **25** 40 min **26** 2500 m **27** \$4.30 **28** 5 kg **29** 20 g
30 \$24 **31** 197 **32** 1.5 L **33** 4000 m^2

UNIT 55 **1** $\frac{3}{8}$ **2** $2\frac{1}{2}$ **3** 24 times **4** $\frac{3}{4}$ and $\frac{1}{4}$ **5** $\frac{2}{5}$ **6** $11\frac{1}{4}$ **7** $7\frac{1}{4}$ **8** 16 halves **9** 10 fifths **10** $6\frac{1}{4}$ hours
11 6 L **12** \$8 **13** 24 **14** \$4.20

UNIT 56 **1** 10 **2** 28 **3** $\frac{3}{25}$ **4** 60 **5** 28 **6** $5\frac{8}{9}$ **7** $\frac{38}{7}$ **8** F **9** $\frac{4}{5}$ **10** $\frac{11}{15}$ **11** $1\frac{10}{63}$ **12** $4\frac{3}{4}$ **13** $\frac{8}{17}$ **14** $5\frac{5}{6}$ **15** $\frac{2}{3}$
16 $2\frac{1}{6}$ **17** \$36 **18** $\frac{1}{9}$ **19** $1\frac{3}{5}$ **20** $\frac{20}{27}$ **21** $\frac{4}{9}$ **22** $6\frac{3}{4}$ **23** 20 **24** $4\frac{3}{10}$ **25** $7\frac{1}{2}$ L **26** 72 **27** $\frac{25}{28}$ **28** $1\frac{1}{9}$ **29** $\frac{4}{9}$ **30** $\frac{5}{9}$

UNIT 57

1

3	$2\frac{1}{2}$	2
$1\frac{1}{2}$	$2\frac{1}{2}$	$3\frac{1}{2}$
3	$2\frac{1}{2}$	2

2

$3\frac{1}{6}$	4	$3\frac{5}{6}$
$4\frac{1}{3}$	$3\frac{2}{3}$	3
$3\frac{1}{2}$	$3\frac{1}{3}$	$4\frac{1}{6}$

3

5	$2\frac{1}{2}$	3
$1\frac{1}{2}$	$3\frac{1}{2}$	$5\frac{1}{2}$
4	$4\frac{1}{2}$	2

4

$3\frac{3}{4}$	$1\frac{1}{4}$	$1\frac{3}{4}$
$\frac{1}{4}$	$2\frac{1}{4}$	$4\frac{1}{4}$
$2\frac{3}{4}$	$3\frac{1}{4}$	$\frac{3}{4}$

5

$1\frac{2}{3}$	$2\frac{1}{2}$	$2\frac{1}{3}$
$2\frac{5}{6}$	$2\frac{1}{6}$	$1\frac{1}{2}$
2	$1\frac{5}{6}$	$2\frac{2}{3}$

6

$1\frac{1}{2}$	$8\frac{1}{2}$	$3\frac{1}{2}$
$6\frac{1}{2}$	$4\frac{1}{2}$	$2\frac{1}{2}$
$5\frac{1}{2}$	$\frac{1}{2}$	$7\frac{1}{2}$

7

$7\frac{1}{2}$	$2\frac{1}{2}$	$9\frac{1}{2}$
$8\frac{1}{2}$	$6\frac{1}{2}$	$4\frac{1}{2}$
$3\frac{1}{2}$	$10\frac{1}{2}$	$5\frac{1}{2}$

8

$5\frac{2}{3}$	$\frac{2}{3}$	$7\frac{2}{3}$
$6\frac{2}{3}$	$4\frac{2}{3}$	$2\frac{2}{3}$
$1\frac{2}{3}$	$8\frac{2}{3}$	$3\frac{2}{3}$

9

$8\frac{1}{2}$	$3\frac{1}{2}$	$10\frac{1}{2}$
$9\frac{1}{2}$	$7\frac{1}{2}$	$5\frac{1}{2}$
$4\frac{1}{2}$	$11\frac{1}{2}$	$6\frac{1}{2}$

10 $3\frac{1}{4}$ **11** $6\frac{3}{4}$ **12** $1\frac{11}{28}$ **13** $2\frac{7}{8}$ **14** $5\frac{11}{20}$ **15** $5\frac{9}{32}$ **16** $3\frac{4}{7}$ **17** $\frac{11}{16}$ **18** $\frac{1}{4}$ **19** $\frac{11}{12}$

UNIT 59

1 a 1 **b** 3 **c** 5 **d** 8 **2 a** 12 **b** 6 **c** 36 **d** 450 **3 a** $\frac{25}{36}$ **b** $2\frac{1}{4}$ **c** $\frac{49}{64}$ **d** $10\frac{9}{16}$
4 a $\frac{5}{8}$ **b** $\frac{2}{3}$ **c** $\frac{7}{11}$ **d** $\frac{11}{13}$ **5 a** $2\frac{3}{5}$ **b** $6\frac{1}{3}$ **c** $9\frac{23}{40}$ **d** $7\frac{15}{28}$ **6 a** 48 **b** \$24 **c** 324 kg **d** 32 h **e** 39 km **f** \$54.40
7 a $1\frac{2}{15}$ **b** 15 **c** $9\frac{1}{3}$ **d** $2\frac{1}{4}$

UNIT 60

1 0.3 **2** 0.1 **3** 0.9 **4** 0.2 **5** 0.13 **6** 0.29 **7** 0.83 **8** 0.91 **9** 0.513 **10** 0.625 **11** 0.334 **12** 0.576 **13** 1.2 **14** 1.7 **15** 5.9 **16** 8.3 **17** 1.27 **18** 3.61 **19** 5.21 **20** 9.63 **21** 4.637 **22** 8.934 **23** 6.537 **24** 9.688 **25** 0.63 **26** 0.59 **27** 0.13 **28** 0.97 **29** 2.89 **30** 2.73 **31** 0.567 **32** 0.936 **33** 1.69 **34** 5.37 **35** 0.689 **36** 0.8 **37** 0.886 **38** 0.669 **39** 0.561 **40** 0.397 **41** 4.43 **42** 0.923 **43** 0.369 **44** 5.68

UNIT 61

1 0.6 **2** 0.52 **3** 0.36 **4** 0.75 **5** 0.55 **6** 0.54 **7** 0.46 **8** 0.68 **9** 0.375 **10** 0.075 **11** 0.815 **12** 0.425 **13** 0.095 **14** 0.95 **15** 0.965 **16** 0.66 **17** 0.44 **18** 0.345 **19** 0.938 **20** 0.766 **21** 0.486 **22** 0.968 **23** 0.78 **24** 0.875 **25** 0.546 **26** 0.405 **27** 0.605 **28** 0.82 **29** 0.664 **30** 3.35

UNIT 62

1 $\frac{1}{10}$ **2** $\frac{3}{10}$ **3** $\frac{7}{10}$ **4** $1\frac{9}{10}$ **5** $\frac{11}{100}$ **6** $\frac{13}{100}$ **7** $\frac{17}{100}$ **8** $\frac{19}{100}$ **9** $\frac{21}{100}$ **10** $\frac{23}{100}$ **11** $\frac{27}{100}$ **12** $\frac{29}{100}$ **13** $2\frac{31}{100}$ **14** $\frac{33}{100}$ **15** $5\frac{37}{100}$ **16** $\frac{39}{100}$ **17** $2\frac{41}{100}$ **18** $\frac{43}{100}$ **19** $3\frac{47}{100}$ **20** $\frac{53}{100}$ **21** $\frac{59}{100}$ **22** $7\frac{61}{100}$ **23** $\frac{71}{100}$ **24** $\frac{73}{100}$ **25** $5\frac{79}{100}$ **26** $\frac{83}{100}$ **27** $6\frac{89}{100}$ **28** $\frac{91}{100}$ **29** $7\frac{93}{100}$ **30** $\frac{97}{100}$ **31** $9\frac{99}{100}$ **32** $\frac{101}{1000}$ **33** $\frac{103}{1000}$ **34** $2\frac{121}{1000}$ **35** $5\frac{321}{1000}$ **36** $8\frac{129}{1000}$ **37** $10\frac{519}{1000}$ **38** $8\frac{213}{1000}$ **39** $9\frac{353}{1000}$ **40** $5\frac{729}{1000}$

UNIT 63

1 $\frac{4}{5}$ **2** $\frac{2}{5}$ **3** $\frac{3}{5}$ **4** $\frac{1}{2}$ **5** $\frac{3}{20}$ **6** $\frac{1}{4}$ **7** $\frac{1}{20}$ **8** $\frac{4}{25}$ **9** $\frac{6}{25}$ **10** $1\frac{1}{4}$ **11** $2\frac{1}{4}$ **12** $3\frac{1}{4}$ **13** $\frac{16}{25}$ **14** $\frac{17}{20}$ **15** $\frac{18}{25}$ **16** $\frac{17}{25}$ **17** $\frac{13}{20}$ **18** $\frac{3}{4}$ **19** $8\frac{9}{20}$ **20** $3\frac{6}{25}$ **21** $\frac{1}{25}$ **22** $\frac{24}{25}$ **23** $\frac{22}{25}$ **24** $1\frac{3}{25}$ **25** $3\frac{1}{2}$ **26** $8\frac{2}{5}$ **27** $7\frac{1}{5}$ **28** $2\frac{3}{4}$ **29** $5\frac{1}{4}$ **30** $\frac{33}{50}$ **31** $3\frac{1}{5}$ **32** $5\frac{3}{5}$ **33** $7\frac{7}{20}$

UNIT 64

1 50% **2** 20% **3** 25% **4** 40% **5** 75% **6** $12\frac{1}{2}\%$ **7** $87\frac{1}{2}\%$ **8** 5% **9** 2% **10** 15% **11** 150% **12** 225% **13** 7% **14** 9% **15** 11% **16** 34% **17** 12% **18** 92% **19** 98% **20** 30% **21** 69% **22** 81% **23** 33% **24** $41\frac{1}{2}\%$ **25** $94\frac{1}{2}\%$ **26** 54% **27** 68% **28** 94% **29** 45% **30** $17\frac{1}{2}\%$ **31** 16.2% **32** 53.9% **33** 63.7%

UNIT 65 **1** $\frac{13}{100}$ **2** $\frac{29}{100}$ **3** $\frac{69}{100}$ **4** $\frac{33}{100}$ **5** $\frac{9}{100}$ **6** $\frac{27}{100}$ **7** $\frac{2}{5}$ **8** 1 **9** $\frac{16}{25}$ **10** $\frac{1}{10}$ **11** $\frac{4}{5}$ **12** $\frac{1}{2}$ **13** $\frac{22}{25}$ **14** $\frac{1}{4}$ **15** $1\frac{1}{2}$ **16** $\frac{3}{4}$ **17** $\frac{3}{20}$ **18** $\frac{7}{100}$ **19** $\frac{17}{100}$ **20** $\frac{24}{25}$ **21** $2\frac{1}{4}$ **22** $6\frac{7}{20}$ **23** $6\frac{3}{5}$ **24** $5\frac{1}{10}$ **25** $\frac{9}{10}$ **26** $1\frac{4}{5}$ **27** $1\frac{9}{25}$ **28** $2\frac{7}{10}$ **29** $5\frac{3}{25}$ **30** $3\frac{3}{5}$ **31** $8\frac{1}{5}$ **32** $3\frac{16}{25}$ **33** $1\frac{1}{5}$

UNIT 66 **1** 75% **2** 50% **3** 25% **4** 23% **5** 15% **6** 83% **7** 33% **8** 10% **9** 20% **10** 12% **11** 19% **12** 31% **13** 80% **14** $12\frac{1}{2}$% **15** 145% **16** 30% **17** 50% **18** 2% **19** 238% **20** 85% **21** 96% **22** 10% **23** 6% **24** 9% **25** 28% **26** 55% **27** 67% **28** 40% **29** 87% **30** 96% **31** 53% **32** 89% **33** 82.9% **34** 29.6% **35** $63\frac{1}{2}$% **36** 58% **37** 61% **38** 21.5% **39** 39% **40** 3.6% **41** 9.1% **42** 90.5% **43** 7% **44** 11% **45** 0.1% **46** 23.7% **47** 38.6% **48** 61.3% **49** 25.2% **50** 230% **51** 894% **52** 16.7% **53** 29.8% **54** 96.6% **55** $34\frac{1}{2}$% **56** 51.4% **57** 69.1% **58** 392% **59** 836% **60** 265% **61** 69.7% **62** 35.7% **63** 12.3% **64** 32.1%

UNIT 67 **1** 0.25 **2** 0.2 **3** 0.1 **4** 0.24 **5** 0.53 **6** 0.27 **7** 0.6 **8** 0.87 **9** 0.99 **10** 0.5 **11** 1 **12** 0.75 **13** 0.625 **14** 0.55 **15** 0.783 **16** 0.635 **17** 0.3 **18** 0.4 **19** 0.65 **20** 0.383 **21** 0.396 **22** 0.92 **23** 0.59 **24** 0.115 **25** 0.63 **26** 0.253 **27** 0.249 **28** 0.365 **29** 0.453 **30** 0.025 **31** 0.68 **32** 0.28 **33** 0.0003 **34** 0.157 **35** 0.738 **36** 0.963

UNIT 68 **1** 0.4, 40% **2** $\frac{1}{20}$, 0.05 **3** $\frac{17}{20}$, 85% **4** 0.3, 30% **5** $\frac{1}{10}$, 0.1 **6** $\frac{1}{8}$, 12.5% **7** 0.28, 28% **8** $\frac{1}{5}$, 0.2 **9** 0.8, 80% **10** $\frac{2}{5}$, 40% **11** $\frac{1}{4}$, 0.25 **12** 0.45, 45% **13** $\frac{4}{5}$, 80% **14** $\frac{1}{2}$, 0.5 **15** $\frac{18}{25}$, 72% **16** 0.375, $37\frac{1}{2}$% **17** $\frac{3}{4}$, 0.75 **18** $\frac{11}{20}$, 55% **19** 0.95, 95% **20** 1, 1 **21** $\frac{17}{20}$, 85% **22** 0.125, $12\frac{1}{2}$% **23** $1\frac{1}{5}$, 1.2 **24** $\frac{13}{20}$, 65% **25** 1.75, 175% **26** $1\frac{1}{2}$, 1.5

UNIT 69 **1** 0.1 **2** 0.2 **3** 0.3 **4** 0.07 **5** 0.4 **6** 0.12 **7** 0.45 **8** 0.06 **9** $\frac{1}{2}$ **10** $\frac{7}{10}$ **11** $\frac{9}{10}$ **12** $\frac{1}{8}$ **13** $\frac{3}{4}$ **14** $1\frac{1}{2}$ **15** $3\frac{3}{10}$ **16** $7\frac{4}{5}$ **17** 90% **18** 12% **19** 7% **20** 75% **21** 83% **22** $33\frac{1}{2}$% **23** 25% **24** 50% **25** $\frac{11}{100}$ **26** $\frac{23}{100}$ **27** $\frac{29}{100}$ **28** $\frac{4}{5}$ **29** $\frac{1}{10}$ **30** $\frac{3}{4}$ **31** $1\frac{1}{5}$ **32** $2\frac{1}{4}$ **33** 25% **34** 37% **35** 81% **36** 93% **37** 7% **38** 250% **39** 55% **40** 365% **41** 0.1 **42** 0.2 **43** 0.35 **44** 0.69 **45** 1.2 **46** 0.125 **47** 0.63 **48** 0.58 **49** 0.283 **50** 0.85 **51** 0.4 **52** 0.6

UNIT 70 **1** 50 **2** $\frac{1}{4}$ **3** 84 **4** 20 **5** 100 **6** $\frac{10}{3}$ **7** $3\frac{6}{11}$ **8** F **9** $\frac{3}{7}$ **10** $\frac{3}{5}$ **11** $\frac{59}{63}$ **12** $5\frac{1}{2}$ **13** $\frac{11}{17}$ **14** $6\frac{6}{7}$ **15** $\frac{1}{8}$ **16** $2\frac{5}{12}$ **17** \$240 **18** $\frac{1}{9}$ **19** $\frac{7}{10}$ **20** $2\frac{1}{3}$ **21** $\frac{15}{22}$ **22** $12\frac{3}{4}$ **23** 81 **24** 32 L **25** 0 **26** $\frac{25}{28}$ **27** $1\frac{7}{50}$ **28** $\frac{4}{25}$ **29** $22\frac{2}{9}$ **30** $\frac{7}{20}$

UNIT 72 **1 a** 0.5 **b** 0.09 **c** 0.23 **d** 0.24 **e** 0.35 **f** 12.3 **2 a** $\frac{9}{10}$ **b** $\frac{13}{100}$ **c** $\frac{1}{8}$ **d** $1\frac{1}{5}$ **e** $30\frac{1}{2}$ **f** $2\frac{9}{25}$ **3 a** 55% **b** 64% **c** 9% **d** 75% **e** 12.5% **f** 125% **4 a** $\frac{29}{100}$ **b** $\frac{3}{5}$ **c** $1\frac{1}{4}$ **d** $\frac{3}{4}$ **e** $\frac{5}{8}$ **f** $\frac{21}{400}$ **5 a** 73% **b** 890% **c** 632% **d** 540% **e** 25% **f** 62.5% **6 a** 0.8 **b** 0.6 **c** 0.55 **d** 1.8 **e** 0.95 **f** 0.33

EXAM PAPER 1

Part A **1 a** $\frac{7}{3}$ **b** $\frac{19}{2}$ **c** $\frac{19}{5}$ **d** $\frac{71}{10}$ **2 a** > **b** > **c** > **d** < **3 a** T **b** F **c** T **d** F **4 a** $\frac{1}{2}$ **b** $\frac{3}{4}$ **c** $\frac{8}{15}$ **d** $\frac{7}{9}$ **5 a** $6\frac{1}{3}$ **b** $7\frac{7}{9}$ **c** $2\frac{2}{5}$ **d** $14\frac{2}{7}$ **6 a** 30 **b** 56 **c** 45 **d** 15 **7 a** $\frac{1}{7}, \frac{1}{5}, \frac{1}{3}$ **b** $\frac{2}{5}, \frac{3}{5}, \frac{2}{3}$ **c** $\frac{1}{4}, \frac{1}{2}, \frac{5}{8}$ **d** $\frac{1}{3}, \frac{3}{8}, \frac{7}{12}$ **e** $\frac{1}{6}, \frac{1}{4}, \frac{1}{3}$ **f** $\frac{1}{2}, \frac{2}{3}, \frac{3}{4}$ **Part B** **8 a** $1\frac{1}{8}$ **b** $\frac{24}{35}$ **c** $6\frac{14}{15}$ **9 a** $\frac{1}{4}$ **b** $\frac{3}{8}$ **c** $1\frac{3}{4}$ **10 a** $\frac{7}{32}$ **b** $\frac{1}{12}$ **c** $17\frac{1}{2}$ **11 a** $10\frac{1}{2}$ **b** $\frac{125}{128}$ **c** $\frac{7}{12}$ **12 a** 40 cm **b** 45 min **c** $\$80$ **d** 162° **e** $\frac{1}{10}$ **f** $\frac{2}{25}$ **g** $\frac{1}{10}$ **h** $\frac{4}{5}$ **13 a** 16 **b** $6\frac{13}{36}$ **c** $5\frac{1}{2}$ **d** 5 **14 a** $\$26$ **b** $\frac{15}{32}$ L **c** 80 km/h **d** $\frac{1}{26}$ **Part C** **15 a** $\frac{9}{16}$ **b** $\frac{25}{64}$ **c** $\frac{16}{81}$ **d** $\frac{49}{100}$ **16 a** $\frac{5}{6}$ **b** $\frac{8}{11}$ **c** $\frac{3}{4}$ **d** $\frac{3}{5}$ **17 a** 0.3 **b** 0.09 **c** 0.6 **d** 0.45 **18 a** $\frac{61}{100}$ **b** $\frac{3}{25}$ **c** $1\frac{1}{4}$ **d** $3\frac{3}{5}$ **19 a** 80% **b** 76% **c** $42\frac{1}{2}\%$ **d** 555% **20 a** $\frac{99}{100}$ **b** $\frac{16}{25}$ **c** $1\frac{1}{4}$ **d** $\frac{3}{10}$ **21 a** 0.07, 7% **b** $\frac{9}{25}$, 36% **c** $\frac{33}{100}$, 0.33

EXAM PAPER 2

Part A **1 a** $\frac{11}{4}$ **b** $\frac{16}{3}$ **c** $\frac{17}{5}$ **d** $\frac{29}{4}$ **2 a** $4\frac{1}{2}$ **b** $3\frac{1}{4}$ **c** $6\frac{3}{7}$ **d** $9\frac{1}{2}$ **3 a** $\frac{5}{7}$ **b** $\frac{5}{7}$ **c** $\frac{2}{3}$ **d** $3\frac{3}{4}$ **4 a** 20 **b** 49 **c** 35 **d** 60 **5 a** < **b** < **c** > **d** < **6 a** T **b** T **c** T **d** F **7 a** $\frac{1}{2}, \frac{1}{5}, \frac{1}{9}$ **b** $\frac{3}{4}, \frac{2}{3}, \frac{3}{5}$ **c** $\frac{3}{4}, \frac{3}{7}, \frac{1}{4}$ **d** $\frac{2}{3}, \frac{5}{8}, \frac{2}{7}$ **e** $\frac{5}{9}, \frac{3}{7}, \frac{4}{11}$ **f** $\frac{2}{5}, \frac{3}{9}, \frac{2}{7}$ **Part B** **8 a** $\frac{3}{5}$ **b** $1\frac{13}{28}$ **c** $11\frac{11}{12}$ **9 a** $\frac{4}{17}$ **b** $\frac{4}{25}$ **c** $4\frac{7}{18}$ **10 a** $\frac{15}{56}$ **b** $\frac{2}{9}$ **c** $78\frac{3}{25}$ **11 a** $1\frac{1}{5}$ **b** $\frac{1}{2}$ **c** $3\frac{1}{3}$ **12 a** 250 m **b** $\$4.30$ **c** 16 kg **d** 197 **e** $\frac{1}{4}$ **f** 320 L **g** $\frac{1}{12}$ **h** $\frac{3}{5}$ **13 a** $3\frac{11}{25}$ **b** $3\frac{47}{108}$ **c** $4\frac{31}{360}$ **d** 42 **14 a** $\frac{3}{5}$ **b** 30 times **c** 20 L **d** $4\frac{1}{6}$ **Part C** **15 a** $\frac{25}{81}$ **b** $\frac{9}{49}$ **c** $\frac{16}{121}$ **d** $\frac{81}{169}$ **16 a** $\frac{5}{9}$ **b** $\frac{7}{13}$ **c** $\frac{5}{6}$ **d** $\frac{3}{8}$ **17 a** 0.7 **b** 0.11 **c** 0.36 **d** 0.26 **18 a** $\frac{23}{100}$ **b** $\frac{18}{25}$ **c** $2\frac{3}{4}$ **d** $5\frac{3}{5}$ **19 a** 10% **b** 85% **c** $24\frac{1}{6}\%$ **d** 450% **20 a** $\frac{97}{100}$ **b** $\frac{43}{50}$ **c** $\frac{33}{100}$ **d** $\frac{3}{50}$ **21 a** 0.25, 25% **b** $\frac{1}{8}$, $12\frac{1}{2}\%$ **c** $\frac{7}{10}$, 0.7

EXAM PAPER 3

Part A **1 a** $\frac{9}{2}$ **b** $\frac{43}{8}$ **c** $\frac{88}{9}$ **d** $\frac{31}{8}$ **2 a** $10\frac{10}{11}$ **b** $22\frac{2}{9}$ **c** $13\frac{5}{7}$ **d** $16\frac{9}{10}$ **3 a** > **b** < **c** < **d** = **4 a** F **b** F **c** T **d** F **5 a** $\frac{7}{11}$ **b** $\frac{1}{25}$ **c** $\frac{3}{4}$ **d** $\frac{2}{3}$ **6 a** 144 **b** 117 **c** 45 **d** 120 **7 a** $\frac{1}{9}, \frac{2}{3}, \frac{5}{7}$ **b** $\frac{5}{12}, \frac{8}{11}, \frac{3}{4}$ **c** $\frac{8}{20}, \frac{12}{15}, \frac{9}{11}$ **d** $\frac{1}{7}, \frac{3}{18}, \frac{10}{25}$ **e** $\frac{5}{17}, \frac{23}{34}, \frac{16}{17}$ **f** $\frac{5}{18}, \frac{3}{7}, \frac{8}{14}$ **Part B** **8 a** $\frac{2}{5}$ **b** $1\frac{13}{55}$ **c** $12\frac{7}{12}$ **9 a** $\frac{2}{7}$ **b** $-\frac{6}{35}$ **c** $2\frac{1}{9}$ **10 a** $\frac{1}{2}$ **b** $\frac{4}{25}$ **c** $20\frac{5}{8}$ **11 a** $\frac{14}{45}$ **b** 1 **c** 5 **12 a** 3 h **b** $3\frac{3}{4}$ km **c** $\$800$ **d** 32 **e** $\frac{3}{40}$ **f** $\frac{1}{20}$ **g** $\frac{1}{9}$ **h** $\frac{1}{50}$ **13 a** $\frac{7}{20}$ **b** 4 **c** $5\frac{1}{4}$ **d** $1\frac{23}{28}$ **14 a** $\frac{2}{5}$ **b** $7\frac{1}{4}$ **c** $16\frac{11}{12}$ **d** $\$333.33$ **Part C** **15 a** $\frac{49}{81}$ **b** $\frac{36}{121}$ **c** $\frac{9}{64}$ **d** $\frac{81}{121}$ **16 a** $\frac{3}{4}$ **b** $\frac{6}{7}$ **c** $\frac{7}{8}$ **d** $\frac{9}{11}$ **17 a** 0.07 **b** 0.125 **c** 0.008 **d** 0.35 **18 a** $\frac{67}{100}$ **b** $\frac{3}{20}$ **c** $3\frac{3}{4}$ **d** $8\frac{1}{2}$ **19 a** 50% **b** 45% **c** 68% **d** 350% **20 a** $\frac{18}{25}$ **b** $\frac{37}{100}$ **c** $1\frac{1}{4}$ **d** $5\frac{1}{2}$ **21 a** 0.8, 80% **b** $\frac{13}{20}$, 65% **c** $\frac{22}{25}$, 0.88

EXAM PAPER 4

Part A **1 a** $\frac{31}{2}$ **b** $\frac{51}{4}$ **c** $\frac{103}{12}$ **d** $\frac{65}{6}$ **2 a** $9\frac{2}{3}$ **b** $8\frac{1}{4}$ **c** $8\frac{3}{7}$ **d** $24\frac{3}{5}$ **3 a** < **b** > **c** = **d** < **4 a** T **b** F **c** T **d** F **5 a** $\frac{8}{9}$ **b** $\frac{5}{8}$ **c** $\frac{8}{11}$ **d** $\frac{4}{15}$ **6 a** 5 **b** 72 **c** 91 **d** 126 **7 a** $\frac{8}{9}, \frac{5}{6}, \frac{2}{3}$ **b** $\frac{3}{7}, \frac{5}{14}, \frac{7}{21}$ **c** $\frac{9}{8}, \frac{8}{9}, \frac{8}{19}$ **d** $\frac{4}{3}, \frac{9}{12}, \frac{5}{7}$ **e** $\frac{60}{65}, \frac{35}{40}, \frac{38}{76}$ **f** $\frac{28}{29}, \frac{23}{25}, \frac{11}{20}$ **Part B** **8 a** $\frac{11}{15}$ **b** $1\frac{5}{24}$ **c** $9\frac{13}{24}$ **9 a** $\frac{3}{14}$ **b** $\frac{2}{9}$ **c** $2\frac{5}{8}$ **10 a** $\frac{35}{36}$ **b** $\frac{2}{5}$ **c** $15\frac{3}{16}$ **11 a** $1\frac{17}{21}$ **b** $2\frac{1}{2}$ **c** $6\frac{1}{6}$ **12 a** 52 **b** 19 min **c** $\$25$ **d** 300 **e** $\frac{1}{17}$ **f** $\frac{1}{20}$ **g** $\frac{1}{30}$ **h** $\frac{3}{5}$ **13 a** $3\frac{71}{72}$ **b** $3\frac{7}{48}$ **c** $\frac{3}{88}$ **d** $\frac{8}{81}$ **14 a** 8 **b** 30 **c** 6 h **d** 2 h 52 min 48 seconds **Part C** **15 a** $\frac{144}{625}$ **b** $\frac{4}{9}$ **c** $\frac{25}{169}$ **d** $39\frac{1}{16}$ **16 a** $\frac{2}{9}$ **b** $\frac{3}{13}$ **c** $\frac{5}{11}$ **d** $1\frac{2}{3}$ **17 a** 0.5 **b** 0.75 **c** 0.28 **d** 0.625 **18 a** $\frac{21}{100}$ **b** $\frac{17}{20}$ **c** $2\frac{1}{4}$ **d** $4\frac{1}{4}$ **19 a** 9% **b** $21\frac{1}{2}\%$ **c** 16.25% **d** 925% **20 a** $\frac{29}{50}$ **b** $\frac{69}{100}$ **c** $2\frac{1}{2}$ **d** $5\frac{3}{5}$ **21 a** 0.23, 23% **b** $3\frac{1}{4}$, 325% **c** $\frac{49}{50}$, 0.98